56 Advances in Biochemical Engineering Biotechnology

Managing Editor: T. Scheper

W0245867

Springer-Verlag Berlin Heidelberg GmbH

Biotreatment, Downstream Processing and Modelling

With Contributions by
P. Bajpai, P. K. Bajpai, D. Dochain, N. N. Dutta,
A. C. Ghosh, R. K. Mathur, A. Mukhopadhyay,
M. Perrier, P. L. Rogers, H. S. Shin, B. Wang

With 41 Figures and 32 Tables

 Springer

ISBN 978-3-662-14781-8   ISBN 978-3-540-68670-5 (eBook)
DOI 10.1007/978-3-540-68670-5

Library of Congress Catalog Card Number 72-152360

This work is subject to copyright. All rights are reserved, whether the whole or part of the material is concerned, specifically the rights of translation, reprinting, re-use of illustrations, recitation, broadcasting, reproduction on microfilms or in other ways, and storage in data banks. Duplication of this publication or parts thereof is only permitted under the provisions of the German Copyright Law of September 9, 1965, in its current version, and a copyright fee must always be paid.

© Springer-Verlag Berlin Heidelberg 1997

Originally published by Springer-Verlag Berlin Heidelberg New York in 1997
Softcover reprint of the hardcover 1st edition 1997

The use of registered names, trademarks, etc. in this publication does not imply, even in the absence of a specific statement, that such names are exempt from the relevant protective laws and regulations and therefore free for general use.

Typesetting: Macmillan India Ltd., Bangalore-25
SPIN: 10474146   02/3020 - 5 4 3 2 1 0 - Printed on acid-free paper

Managing Editor

Professor Dr. T. Scheper
Institute of Technical Chemistry, University of Hannover
Callinstraße 3, D - 30167 Hannover, FRG

Editorial Board

Prof. Dr. *W. Babel*   Center of Environmental Research
                                Leipzig-Halle GmbH
                                Section of Environmental Microbiology
                                Peermoserstraße 15
                                D-04318 Leipzig/FRG

Prof. Dr. *H.W. Blanch*   University of California
                                Department of Chemical Engineering
                                Berkely, CA 94720-9989/USA

Prof. Dr. *Ch. L. Cooney*   Massachusetts Institute of Technology
                                Department of Chemical Engineering
                                25 Ames Street
                                Cambridge MA 02139/USA

Prof. Dr. *S.-O. Enfors*   Department of Biochemistry and Biotechnology
                                Royal Institute of Technology
                                Teknikringen 34, S - 100 44 Stockholm/Sweden

Prof. Dr. *K.-E. L. Eriksson*   Department of Biochemistry
                                A214 Life Science Building
                                The University of Georgia
                                Athens, GA 30602-7229/USA

Prof. Dr. *A. Fiechter*   Institut für Biotechnologie
                                Eidgenössische Technische Hochschule
                                ETH-Hönggerberg, CH-8093 Zürich/Switzerland

Prof. Dr. *A. M. Klibanov*   Massachusetts Institute of Technology
                                Department of Chemistry
                                Cambridge, MA 02139/USA

Prof. Dr. *B. Mattiasson*   Department of Biotechnology
                                Chemical Center, Lund University
                                P.O. Box 124, S - 221 00 Lund/Sweden

Prof. Dr. *S. B. Primrose*   21 Amersham Road
                                High Wicombe, Bucks HP13 6QS/U.K.

Editorial Board

Prof. Dr. *H. J. Rehm*   Westfälische Wilhelms Universität
                                Institut für Mikrobiologie
                                Corrensstr. 3, D - 48149 Münster/FRG

Prof. Dr. *P. L. Rogers*   Department of Biotechnology, Faculty of Applied
                                Science, The University of New South Wales
                                Sydney 2052/ Australia

Prof. Dr. *H. Sahm*   Institut für Biotechnologie
                                Forschungszentrum Jülich GmbH, D - 52428 Jülich/FRG

Prof. Dr. *K. Schügerl*   Institut für technische Chemie, Universität Hannover
                                Callinstr. 3, D-30167 Hannover/FRG

Prof. Dr. *G. T. Tsao*   Director, Lab. of Renewable Resources Eng.
                                A. A. Potter Eng. Center, Purdue University
                                West Lafayette, IN 47907/USA

Dr. *K. Venkat*   Phyton Inc., 125 Langmuir Lab
                                95 Brown Road, Ithaca, NY 14850-1257/USA

Prof. Dr. *John Villadsen*   Instituttet for Bioteknologi, Dept. of Biotechnology
                                Technical University of Denmark
                                Bygning 223, DK-2800 Lyngby/Denmark

Prof. Dr. *U. von Stockar*   Institut de Génie Chimique
                                Département de Chimie
                                Ecole Polytechnique Fédérale de Lausanne
                                CH - 1015 Lausanne/Switzerland

Prof. Dr. *C. Wandrey*   Institut für Biotechnologie
                                Forschungszentrum Jülich GmbH
                                Postfach 1913, D - 52428 Jülich/FRG

Attention all "Enzyme Handbook" Users:

A file with the complete volume indexes Vols. 1 through 11 in delimited ASCII format is available for downloading at no charge from the Springer EARN mailbox. Delimited ASCII format can be imported into most databanks.

The file has been compressed using the popular shareware program "PKZIP" (Trademark of PKware INc., PKZIP is available from most BBS and shareware distributors).

This file distributed without any expressed or implied warranty.

To receive this file send an e-mail message to:
SVSERV@DHDSPRI6.BITNET.
The message must be: "GET/ENZHB/ENZ_HB.ZIP".

SPSERV is an automatic data distribution system. It responds to your message. The following commands are available:

HELP	returns a detailed instruction set for the use of SVSERV,
DIR (*name*)	returns a list of files available in the directory "name",
INDEX (*name*)	same as "DIR"
CD <*name*>	changes to directory "name",
SEND <*filename*>	invokes a message with the file "filename"
GET <*filename*>	same as "SEND".

Table of Contents

**Realities and Trends in Enzymatic Prebleaching
of Kraft Pulp**
P. Bajpai, P. K. Bajpai 1

Biotransformation for *L*-Ephedrine Production
P. L. Rogers, H. S. Shin, B. Wang 33

**Inclusion Bodies and Purification of Proteins
in Biologically Active Forms**
A. Mukhopadhyay 61

Extraction and Purification of Cephalosporin Antibiotics
A. C. Ghosh, R. K. Mathur, N. N. Dutta 111

**Dynamical Modelling, Analysis, Monitoring and Control
Design for Nonlinear Bioprocesses**
D. Dochain, M. Perrier................................. 147

Author Index Volumes 51 - 56 199

Subject Index 203

Realities and Trends in Enzymatic Prebleaching of Kraft Pulp

Pratima Bajpai and Pramod K. Bajpai
Chemical Engineering Division,
Thapar Corporate Research & Development Centre,
Patiala-147001, India

List of Abbreviations 2
1 Introduction 3
2 Origin of Enzymes in Bleaching 5
3 The Action of Xylanases on Pulp 7
4 Production of Xylanases for Bleaching 7
5 Factors Affecting the Performance of the Enzymes. 8
   5.1. Effects of Process Conditions on Enzyme Performance 8
       5.1.1 PH 8
       5.1.2 Temperature 9
       5.1.3 Enzyme Dispersion 9
       5.1.4 Reaction Time. 10
   5.2 Effects of Mill Operations on Enzyme Performance 10
       5.2.1 Raw Material 10
       5.2.2 Pulping Process. 10
       5.2.3 Brown Stock Washing. 11
       5.2.4 Bleaching Sequence. 11
6 Enzyme Treatment in Mills 12
7 Effect of Xylanase on Conventional and Unconventional Bleaching 12
   7.1 Lab Trials with Xylanases 13
   7.2 Plant Scale Trials with Xylanases. 22
8 Benefits from Xylanase Treatment 25
9 Future Developments 26
10 Conclusions. 27
11 References 27

Use of biotechnology in pulp bleaching has attracted considerable attention and achieved interesting results in recent years. Enzymes of the hemicellulolytic type, particularly xylan-attacking enzymes, xylanases, are now used in commercial mills for pulp treatment and subsequent incorporation into bleach sequences. There are various reasons for mills to consider the use of bleaching enzymes. Some of the primary reasons are environmental (e.g. reductions in chlorine, chlorine dioxide, and hypochlorite) or economic (decreased chlorine dioxide and/or peroxide requirement), or relate to improved pulp quality (higher brightness ceiling) and improved mill flexibility.
   Although environmental pressures on the pulp producers were responsible for the initial interest in new technologies or biochemical solutions for eliminating chlorine-containing chemicals, which may still be the case in certain parts of the world, there is now a consumer-led demand for elemental chlorine-free (ECF) and total chlorine-free (TCF) pulps. ECF and TCF pulp production offer opportunities for enzymes, which provide a simple and cost-effective way to reduce the use of bleaching chemicals. Enzymes also offer an approach to addressing the need for the elimination of bleach plant effluent discharge. The current developments in enzymatic prebleaching are reviewed here within an engineering context.

Advances in Biochemical Engineering
Biotechnology, Vol. 56
Managing Editor: Th. Scheper
© Springer-Verlag Berlin Heidelberg 1997

List of Abbreviations

AOX   Adsorbable organic halogens
BOD   Biochemical oxygen demand
COD   Chemical oxygen demand
DP   Degree of polymerization
DMSO   Dimethyl sulphoxide
ECF   Elemental chlorine free
ISO   International standard organization
MCC   Modified continuous cooking
PV   Photovolt
TCF   Total chlorine free
TOCl   Total organic chlorine

Bleaching stages

C   Chlorination
C_D Chlorination, with addition of a small amount of chlorine dioxide
(C + D)   Chlorination, with a mixture of chlorine and chlorine dioxide with chlorine in excess
C/D   Sequential bleaching, with chlorine and subsequent treatment with chlorine dioxide without washing between the addition of chemicals
D   Chlorine dioxide treatment
D_c Chlorine dioxide treatment with addition of a small amount of chlorine
D/C   Sequential bleaching, with chlorine dioxide and subsequent chlorination without washing between the addition of chemicals
E   Alkaline extraction
E_H Alkaline extraction in the presence of hypochlorite
E_0 Alkaline extraction in the presence of oxygen
E_{OP} Alkaline extraction in the presence of oxygen and hydrogen peroxide
E_P Alkaline extraction in the presence of hydrogen peroxide
H   Hypochlorite treatment
O   Oxygen delignification
P   Hydrogen peroxide treatment
Q   Chelation of metals
X   Enzyme treatment
Z   Ozone treatment

1 Introduction

A vast pulp and paper industry exists around the world to supply an ever increasing demand for a wide variety of paper products. The kraft process is the world's major pulping method and is likely to remain so in the foreseeable future. It has evolved over a period of 100 years and has became highly refined. Currently more than 70% of the world's annual pulp output of approximately 100 million tonnes is produced by the kraft process. Despite some shortcomings, it is the most cost-effective, versatile and efficient wood delignification method available. Because of this fact and the large amount of capital already invested in kraft pulping, it is unlikely that the process will be replaced in the near future.

The kraft process results in the degradation and solubilization of lignin. Wood chips are cooked in a solution of $Na_2S/NaOH$ at about $170\,°C$ for about 2 h to degrade and solubilize the lignin [1, 2]. The lignin undergoes a variety of reactions but the most important of these results in partial depolymerization and formation of ionizable (mainly phenolic hydroxyl) groups; these changes lead to the dissolution of lignin in the alkaline pulping liquor. The lignin reactions involved in kraft pulping have been studied extensively [3–6]. About 90% of the lignin is removed, the 10% or so remaining in the pulp is primarily responsible for the brown colour characteristic of kraft pulp and papers. The lignin remaining in the pulp has been heavily modified and its persistance probably reflects covalent binding to the hemicelluloses [7]. The brown colour is due to various conjugated structures including quinones, complexed catechols, chalcones and stilbenes, which absorb visible light [3, 8].

Almost half of the kraft pulp produced annually is bleached before use. Because the bleaching process is costly and results in some weight loss, bleached kraft pulp sells for about 10–20% more than unbleached pulp. Bleaching is done by processes employing mainly chlorine and its oxides. The residual lignin is degraded and dissolved in various sequences of bleaching and extraction steps in which chlorine, hypochlorite, chlorine dioxide, oxygen and hydrogen peroxide are used. The use of chlorine as a bleaching agent is a cause for environmental concern since the process produces dioxins and other organochlorine compounds which contribute to the discharge of AOX (adsorbable organic halogens) into receiving water. Gaseous chlorine and hypochlorite are blamed for the formation of chlorinated organic compounds [9, 10].

The conventional way of bleaching kraft pulp has developed during the years and currently the most likely sequence is a C/DEDED for softwood and DEDED for hardwood, although a short version, DEDD, is used by modern mills. This means that most of the hardwood bleach sequences are already elemental chlorine free (ECF), and as the degree of substitution in softwood bleaching is also increasing and approaching 100% (Fig. 1) [11], the kraft mills will soon produce mainly ECF pulps. Also, the demand for total chlorine-free (TCF) pulps is expected to increase but at the moment, there is considerable debate and no clear agreement on a forecast for the rate of growth or the

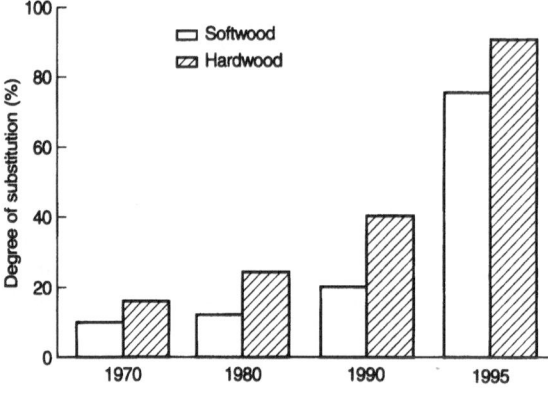

Fig. 1. Replacement of C with D in the first chlorination stage [11]

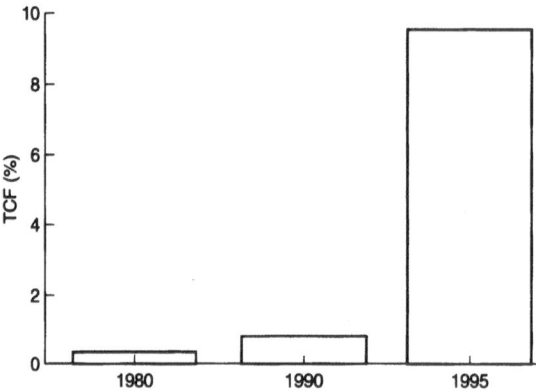

Fig. 2. Proportion of TCF in total pulp production [11, 12]

ultimate size of the TCF market. It was estimated [12] that 3–3.5 million tons of TCF pulps would be produced in 1993 (10% of kraft pulp) with an annual growth rate of 1–1.5 million tons (Fig. 2) [11]. The chlor-alkali producers also have different views on whether or when a phase-out of chlorine will occur in the pulp and paper industry [13]. In the end, the deciding factors will be the success of environmental groups in their moves against the use of chlorine and consumer demands for chlorine-free paper. The success of pulp and paper industry in informing the general population and overcoming the more emotive issues with clear scientific data will also have an influence. Regardless of the pace of change, a market has been created for ECF and TCF pulps where either alternatives have to be found to chlorine-containing bleaching chemicals or new technologies have to be considered. The options open to pulp mills considering a change to chlorine-free bleaching are shown in Table 1 and most of these involve process modifications and/or capital investment. This climate of change has provided an opportunity for enzymes [14–30]. The use of enzymes is an

Table 1. Alternatives available to produce TCF or ECF pulps

Alternatives	Advantages	Disadvantages
Susbstitution of chlorine dioxide for chlorine	Lower AOX	High bleaching costs High investment may be required to satisfy increased demand for chlorine dioxide
Oxygen delignification to reduce kappa prior to bleaching	Lower AOX Lower bleaching costs	High investment
Hydrogen peroxide to replace chlorine-based chemicals	Lower AOX	High bleaching costs Risk of pulp viscosity and strength
Ozone to replace chlorine-based chemicals	Lower AOX	Very high investment Risk of pulp viscosity and strength
Extended cooking to reduce the kappa number before bleaching	Lower AOX Lower bleaching costs	High investment
Enzymes	Lower AOX Reduced use of bleaching chemicals Minimal capital investment Improved strength and brightness	Chances of reduction in yield due to some loss of hemicellulose

innovative answer to chlorine-reduced and chlorine-free bleaching of kraft pulp. In this article, an overview of the recent developments in the application of enzymes in kraft pulp bleaching is presented. The results of international research on the use of enzymes in bleaching, the opportunities for using enzymes, the impact of enzymes on bleaching performance and the contribution of enzymes to the economics of pulp bleaching are discussed. Future prospects/developments in this area are also discussed. Because this is a rapidly emerging field, much of the literature is found in symposium proceedings rather than in the peer-reviewed journals.

2 The Origin of Enzymes in Bleaching

The enzymes used commercially in pulp bleaching are hemicellulases, which selectively affect the accessible hemicellulose fraction of wood pulps. A number of enzymes have been studied but xylanases have been shown to be most effective. The concept of biological bleaching with xylanase emerged from efforts to selectively remove hemicellulose from chemical pulps to produce cellulose acetate [31]. At approximately the same time, a research program jointly carried

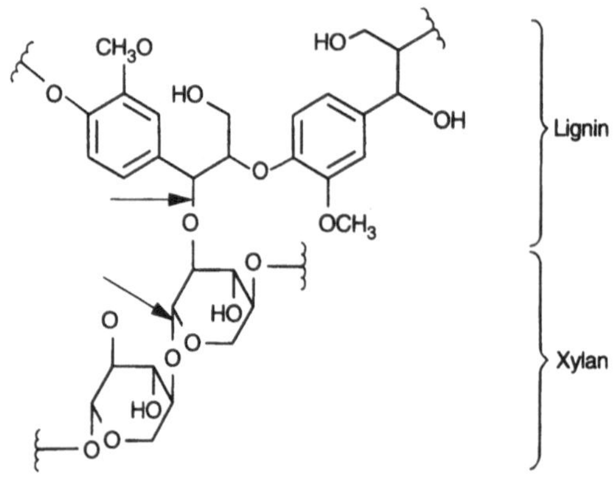

Fig. 3. Possible structure of lignin-carbohydrate complexes in unbleached kraft pulp

out by the Finnish Pulp and Paper Research Institute and Technical Research Centre of Finland was focusing on lignin degrading biochemical processes. It was found that treatment of chemical pulps with xylanases leads to savings in the consumption of bleaching chemicals, decreased environmental loadings or increased final brightness of pulp [32–34]. Since then various papers have been published describing the benefits of xylanase treatments in pulp bleaching. These benefits are environmental (e.g. reductions in chlorine, chlorine dioxide, hypochlorite), economic (decreased chlorine dioxide and/or peroxide requirements), improved pulp quality (higher brightness ceiling) and improved mill flexibility. Mill trials began as early as 1989 in Finland and since 1991, commercial use of xylanase has become a reality. As reported by Jurasek and Paice [35] at the International Symposium on Pollution Prevention in the Manufacture of Pulp and Paper, 10 mills were said to use xylanase prebleaching on a commercial basis and more than 80 mill trials were carried out. In 1993, the process gained even wider acceptance, especially in Canada and Europe. The factors explaining this rapid development are many but can be summarized as follows:

- Xylanase prebleaching belongs to the soft technologies that require very little or no capital investment to operate.
- Process changes are minimal in most cases (neutralization of brown stock). Mill trials are very simple, inexpensive and involve minimal risk.
- Xylanase helps to reduce pollution from bleaching.
- Savings on chemicals can pay for the process.
- Xylanase may help to increase mill capacity where there are chlorine dioxide limitation.
- The process is easily combined with many bleaching sequences for ECF and TCF pulps.

The main enzyme needed to enhance the delignification of kraft pulp is reported to be *endo*-β-xylanase (EC 3.2.1.8) but enrichment of xylanase with other hemicellulolytic enzymes has been shown to improve the effect of enzymatic treatment [34, 36–40]. In the enzymatic pretreatment for bleaching, the hydrolysis of hemicellulose is restricted to a minimum by using only a small amount of enzymes in order to maintain a high pulp yield and the advantageous properties of hemicellulose in pulp [32–33]. Hemicellulose in pulp plays an important role both in fibre morphology and fibre physics. Retention of hemicellulose increases the pulp yield, improves pulp strength and affects fibre quality.

3 The Action of Xylanases on Pulp

The way in which xylanase prebleaching affects subsequent bleaching is not well understood. One possible explanation is that the disruption of the xylan chain by xylanases appears to cleave lignin-carbohydrate bonds (Fig. 3), improves the accessibility of the bleaching chemicals to the pulps and facilitates easier removal of solubilized lignin in bleaching. Paice et al. [40] have shown that there is a significant decrease in xylan DP, and only a small amount of xylan is removed during xylanase prebleaching. The decreased chain length of xylan or its removal results in increased freedom for lignin to diffuse from the hemicellulose-lignin matrix. Another possible explanation involves the role of redeposited xylan [29, 34, 39, 41]. It was known that parts of xylan initially dissolved in kraft cooking liquor could be readsorbed or reprecipitated on and within the pulp fibres. The redeposited xylan may physically shield the residual lignin from bleaching chemicals. Xylanases hydrolyze part of the redeposited xylan, allowing better access of bleaching chemicals to the residual lignin and easier extraction of lignin from pulp fibres. However, this theory is still not conclusive. Pederson et al. [42] found that the xylanase specifically attacks a small fraction of the xylan in pulp fibres and demonstrated that the removal of reprecipitated xylan with dimethylsulphoxide (DMSO) does not improve the bleachability of the pulp. Therefore, he concluded that the DMSO-extractable xylan was not involved in bleach boosting.

4 Production of Xylanases for Bleaching

Several criteria are essential for choosing a microorganism to produce xylanases. In addition to giving the desired biobleaching effect, the resulting enzyme preparation must be produced in sufficiently high quantity and the xylanase technology must be compatible with the technology of a pulp mill. Also, it is very essential that the enzyme preparation should be completely free of cellulase side activity. Any cellulase activity will have serious economic implications in terms of cellulose loss, degraded pulp quality and increased effluent treatment cost. Noncellulolytic preparations have been produced by recombinant DNA technology selective inactivation or bulk scale purification [36, 43, 44].

High productivity has been achieved by exhaustive screening, genetic engineering and growth optimization programs. To produce xylanases, the selected organism is grown for several days in sealed process vessels containing nutrients and oxygen under specific conditions of pH, temperature and agitation. During this time, it secretes enzymes into the growth medium. The living cell mass is then removed, leaving a xylanase rich liquid. This is then concentrated, assayed to determine its activity and packaged for shipment to pulp mills. With the addition of bacteriostatic preservatives, the xylanase preparation remains stable for months. Excessive temperature or freezing can cause loss of activity and should be avoided. The xylanase preparation is not corrosive or reactive and does not need resistant materials for handling.

Table 2 shows the list of some of the commercial xylanases used for prebleaching. The cost of enzyme preparation is in the range of $2–7 t^{-1} pulp processed depending on specific bleaching conditions. Ecopulp XM contains both xylanase and mannanase. The later is targeted at the glucomannan in softwoods.

5 Factors Affecting the Performance of Enzymes

5.1 Effects of Enzyme Process Conditions on Enzyme Performance

Because of the nature of the enzyme/pulp interaction, several factors must be taken into account in order to use enzymes effectively in a mill. Although all of the commercial enzymes act primarily on xylan, the conditions for mill usage are quite different. In addition, to get best value from enzyme usage, the conditions chosen for any one enzyme should be tailored to each mill, based upon careful laboratory testing. The key factors include the pH, temperature, enzyme dispersion and reaction time.

5.1.1 pH

The pH optimum and operating range for enzyme treatment varies among enzymes but generally falls between pH 4 and 8. It is important to note, however, that the optimum pH for a given enzyme may vary among mills, due to different operating conditions described above [45, 46]. Because of this, the optimum pH must be determined for each mill. For some enzymes, the pH

Table 2. Commercial xylanases

Product	Supplier	Approximate price ($/kg)
Pulpzyme HA	Novo Nordisk, Denmark	10
Pulpzyme HB	Novo Nordisk, Denmark	6
Pulpzyme HC	Novo Nordisk, Denmark	n.a.[a]
Irgazyme 10	Genencor, Finland	n.a.
Albazyme 10	Ciba Geigy, Switzerland	
Irgazyme 40-4X	Genencor, Finland	16
Albazyme 40-4X	Ciba Geigy, Switzerland	
Cartazyme HS	Sandoz Chemicals, U.K.	30
Cartazyme HT	Sandoz Chemicals, U.K.	n.a.
Bleachzyme F	Biocon India, Bangalore	10
VAI Xylanase	Voest Alpine, Austria	n.a.
Ecopulp X-200	Alko Ltd. Biotechnology, Finland	7.5
Ecopulp XM	Alko Ltd. Biotechnology, Finland	n.a.
Xylanase	Iogen Corporation, Canada	n.a.
Xylanase L-8000	Solvay Interox, USA	n.a.
Ecozyme	Zeneca Bioproducts ICI, Canada	n.a.

[a] n.a. Not available

optimum spans at least one pH unit, which has proven to be well within the capabilities of pH control of the brown stock. The breadth of pH range is another property that varies among enzymes.

5.1.2 Temperature

The temperature optimum and operating range for enzyme treatment varies among enzymes but is between 35 °C and 60 °C. Cooler temperatures result in similar effects, but over longer treatment times. The maximum operating temperature differs among enzymes. For a given enzyme, the maximum operating temperature varies among mills, mostly due to differences in the extent of brown stock washing [45].

5.1.3 Enzyme Dispersion

The adequate dispersion of enzyme and acid into the pulp is extremely important for enzyme performance. Tracer studies should be conducted at each new installation to assess the adequacy of the dispersion. In general, the degree of mixing depends on the equipment that is used to add enzyme to the pulp and on the absorbency of the brown stock. While medium consistency pumps usually provide adequate mixing of the enzyme into the pulp, the results with thick stock pumps are highly variable. Thick stock pump systems can, however, be configured to approach the mixing performance of a medium consistency pump. The configuration of the optimal system will depend on the specific layout, equipment and metallurgy of the mill.

5.1.4 Reaction Time

A minimum of 1 to 2 h of residence time is required for the enzyme treatment. There is little enzyme action on the pulp beyond 4 to 6 h.

5.2 Effects of Mill Operations on Enzyme Performance

Mill operations also affect the performance of enzymes. The effects of (1) raw material, (2) pulping process, (3) brown stock washing and (4) bleaching sequence should be assessed by laboratory testing prior to mill usage of enzymes.

5.2.1 Raw Material

Among raw materials, the most important distinction is between hardwood and softwood because the respective hemicellulose structures are different. Most of the mill experience has been with softwood. In general, hardwood hemicellulose is more accessible to xylanase enzyme action than is softwood. The magnitude of enzyme benefit is thus greater on hardwood than on softwood. Among hardwood and softwood species, there is some variation in the response to enzyme treatment. However, these differences are not nearly as significant as the effects of the pulping operations.

5.2.2 Pulping Process

The pulping process can affect the content and structure of the hemicellulose in the pulp. This, in turn, changes the extent of enzyme action that is achievable with the pulp. For example, sulfite pulping destroys most of the hemicellulose and thus sulfite pulp is not suitable for enhanced bleaching by enzymatic treatment. Kraft pulping at severe conditions, such as conventional cooking of softwood to kappa number less than 23, also destroys much of the hemicellulose that is accessible to the enzyme. On the other hand, MCC or oxygen-delignifed pulps with low unbleached kappa number respond well to enzyme treatment [45]. Tolan [45] reported a much smaller enzyme benefit for batchcooked pulp at kappa number 21 than for MCC and oxygen-delignified pulp at the same kappa number. The MCC and oxygen-delignified pulps have hemicellulose structures that are similar to that for conventional, high kappa number pulps. Enzyme benefits have been achieved in mills with conventional, MCC and oxygen delignification systems [45, 46].

Table 3. The effect of hemicellulases on the peroxide delignification of unbleached birch and pine kraft pulp

Treatment	Birch kraft pulp			Pine kraft pulp		
	Kappa No.	Brightness (% ISO)	Viscosity (mPa s)	Kappa No.	Brightness (% ISO)	Viscosity (mPa s)
Enzyme from *A. awamori* (500 nkat g^{-1})	8.3	65	8.2	13.1	53	8.3
Enzyme from *A. awamori* (4000 nkat g^{-1})	6.5	72	8.3			
Enzyme from *S. olivochromogenes* (28 nkat g^{-1})	–	–	–	14.4	56	9.2
Buffer[a] washing	9.9	67	8.6	16.2	54	10.0
Reference (peroxide)	12.3	48	8.8	19.7	35	9.4

Birch kraft pulp: Original kappa No. 17.7; viscosity 13.5 mPa s
Pine kraft pulp: Original kappa No. 30.3; viscosity 11.4 mPa s
Based on data from Ref. 32
[a] Citrate buffer

5.2.3 Brown Stock Washing

The brown stock black liquor (brown white water) properties vary greatly from mill to mill. Some mills black liquor can inhibit enzyme performance due to the presence of highly oxidizing compounds. This effect differs significantly among enzymes and should always be checked before proceeding with full scale enzyme use. In those mills where the black liquor is not inhibitory, mill experience, has shown that the day-to-day variation in brown stock washing has little impact on enzyme performance. However, the extent of washing can affect the maximum enzyme treatment temperature, which is important in mills that are run as hot at possible. For example, at 25 kg t^{-1} of soda, the maximum temperature tolerated by Iogen's enzyme is about 5 °C less than for a typical mill-washed pulp. This property varies among enzymes [45].

It is important to note that it is not necessary to wash the pulp after enzyme treatment (before chlorination) to achieve the enhanced bleaching. Identical enzyme benefits have been obtained [47] with and without a post-enzyme washing. This indicates, not surprisingly, that enzyme treatment does not solubilize lignin. However, a post-enzyme washing might be beneficial to some pulps [45, 48].

5.2.4 Bleaching Sequence

The bleaching sequence influences the enzyme's benefit to the mill in several ways. In general, the chemical savings on softwood by enzyme treatment can be taken as any combination of chlorine and chlorine dioxide that comprises about 15% of the total active chlorine used [46]. However, mill experience has confirmed that lab data which showed that the chemical savings depend on the identity and relative amounts of bleaching chemicals used (i.e. chlorine dioxide substitution, peroxide, oxidative extraction) as well as on the relative balance between chemical usage in the delignification. When a mill varies its bleaching sequence, it can expect a change in the enzyme's benefit as well. The specific bleaching sequence should be tested on enzyme treated pulp on a laboratory scale before mill implementation [45].

6 Enzyme Treatment in Mills

Typically the enzyme is added as an aqueous solution to the pulp at the final brown stock washer. Because enzymes are extremely potent catalysts, the desired effects are produced with small amounts of enzymes. The brown stock, which (though washed) is highly alkaline (pH 9–12), must be neutralized with acid, usually sulfuric, to be compatible with enzyme treatment.

The pulp is then pumped to the high density storage tower where the enzyme acts. From there the enzyme treated pulp is then pumped into the first bleaching tower where the first contact with the oxidizing chemicals destroy the enzyme. Unlike other bleaching chemicals, xylanases do not brighten or delignify the pulp. They modify the pulp to make the lignin more accessible to removal by other bleaching chemicals. The enzymatically treated pulp then passes through the bleach plant with a decreased chemical requirement for bleaching.

7 Effect of Xylanase Enzyme on Conventional and Unconventional Bleaching

When a mill wants to either reduce its conventional bleaching costs or improve the economy of TCF bleaching, it is faced with the fact that the existing processing equipment has to be used. The standard process layout for conventional and TCF bleaching is presented in Figs. 4 and 5. The enzyme is added before the brown stock tower. To achieve a good effect, it has to be ensured that mixing is sufficient both after pH adjustment and enzyme addition. In general, it is very easy to arrange for correct addition points for both acids and enzyme. As the amount of chemicals are small, the equipment needed for these chemicals is also minimal. Generally, a permanent acid addition facility is installed but the enzyme can easily be added from the container.

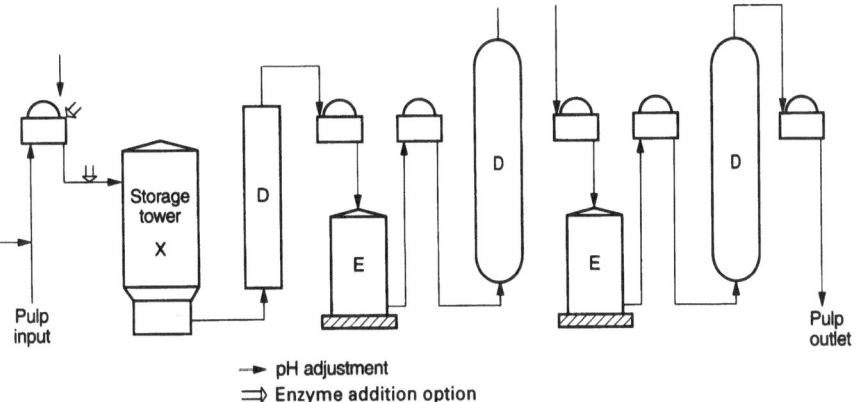

Fig. 4. Xylanase application in conventional bleaching

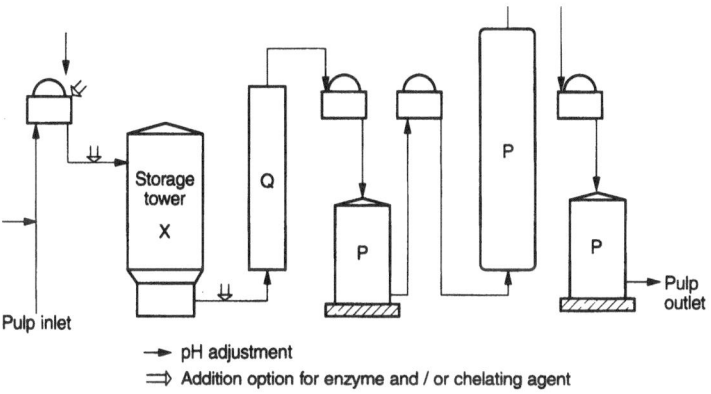

Fig. 5. Xylanase application in TCF bleaching

Results from laboratory studies and mill trials show about 35–41% reduction in active chlorine at the chlorination stage for hardwoods and 10–26% for softwoods, whereas savings in total active chlorine are found to be 20–25% for hardwoods and 10–15% for softwoods, if the pulps are pretreated with xylanase enzyme.

7.1 Lab Trials with Xylanases

Viikari and coworkers [32] reported a 25% reduction in the consumption of active chlorine by the enzymatically pretreated pine kraft pulp or for the same charge of active chlorine, delignification to lower kappa numbers (a measure of residual lignin) than the reference pulps. They also reported a significant reduction in chlorine dioxide consumption by a hemicellulase-treated pine kraft pulp, when it was subsequently bleached to 89–90% ISO brightness by a D_cEDED sequence. Kappa numbers of unbleached pine and birch kraft pulps were both reduced by 50% from original values by treating the pulps with enzyme followed by peroxide, with an increase in brightness concomitant with the reduction in kappa number. These pulps were comparable to chlorine-bleached pulps (Table 3).

Paice et al. [36] and Jurasek and Paice [49] used the pure enzymes produced by Bernier et al. [50] from clones of *Escherichia coli* to remove lignin from kraft pulp. The pulp obtained by combined enzyme and CED bleaching showed improved brightness compared with the conventionally bleached pulp. The final brightness of enzymatically treated pulp was 83.2% which was 3.2 points higher than the control.

Senior et al. [48, 51–54] conducted a systematic study on the effect of xylanase upon C/DEDED sequence using different levels of chlorine dioxide substitution and kappa factor, where kappa factor is the amount of chlorine, in weight percent of pulp, divided by kappa number. They found reduced chemical use and higher brightness for both hardwood pulps and softwood pulps (Tables 4 and 5). The maximum xylanase effectiveness was achieved at low chlorine dioxide substitution for hardwood pulps and at low or high substitution for softwood pulps.

Repligen Sandoz Research Corporation and the Department of Wood and Paper Science at the North Carolina State University developed the xylanase enzyme Cartazyme HS with no contaminating cellulase side activity. The use of Cartazyme helped to achieve chlorine-free bleaching, resulting in a pulp with 89.6% brightness compared to a control sample which achieved a brightness of 83.8% [55].

Clark et al. [56] evaluated the use of hemicellulolytic enzymes for improving the bleachability of radiata pine kraft pulp. Treatment with these enzymes was found to produce savings of 20–25% of chlorine chemicals used during subsequent bleaching. Pederson et al. [57] studied the bleach boosting of kraft pulp with alkaline xylanase preparations completely free of cellulase. These preparations gave good bleach boosting effects at pH 8–9 on kraft pulp. Holm [27] reported enzymatic bleach boosting of Swedish birch kraft pulp delignified with oxygen and of a conventionally cooked *Eucalyptus globulus* with commercial xylanase – Pulpzyme HB. The amount of active chlorine (in the form of chlorine dioxide) required in subsequent bleaching was found to be reduced by 32% and 28%, respectively. Experiments with North American pine kraft also showed the benefits of the enzyme boost process prior to an ozone bleaching sequence. Details of this product and guidelines for its plant scale application were reported by Gibson [58].

Tolan and Canovas [47] found that pulp treated with enzyme in a D/CE_0DED sequence using 50% chlorine dioxide substitution required 16% less total active chlorine to obtain 90% ISO brightness. Dunlop and Gronberg

Table 4. Effect of xylanase treatment on C/DEDED bleaching sequences for various kappa factors and chlorine dioxide substitutions

Pulp	Chlorine dioxide substitution (%)	Kappa factor	Final brightness (% ISO)
Control	10	0.200	90.0
		0.233	90.2
		0.266	90.3
	40	0.150	89.2
		0.173	90.0
		0.200	91.0
	70	0.200	90.0
		0.233	90.6
		0.250	90.9
	100	0.150	83.5
		0.175	85.3
		0.200	87.0
Xylanase pretreated	10	0.05	90.0
		0.10	90.8
		0.15	91.8
	40	0.05	90.0
		0.10	90.8
		0.15	91.8
	70	0.05	88.7
		0.10	90.0
		0.15	90.3
	100	0.10	87.4
		0.15	87.4

Based on data from Ref. 51

Table 5. Total chlorine and chlorine dioxide charges needed to achieve 90% ISO brightness for xylanase pretreated and untreated hardwood pulps

Pulp	Chlorine dioxide substitution (%)	Total chlorine charge on pulp (%)
Control	10	6.75
	40	6.25
	70	7.00
Xylanase pretreated	10	4.55
	40	4.70
	70	5.05

Based on data from Ref. 52

[11] reported about 20–25% savings in active chlorine with hardwoods and 10–15% with softwoods using commercial enzyme Cartazyme HS-10.

Bajpai et al. [59, 60] reported that pretreatment of eucalyptus kraft pulp with commercial xylanases – Pulpzyme HA, Novozyme 473 and VAI xylanase – in a CEH bleaching sequence, resulted in a 31% reduction in chlorine consumption at the chlorination stage. Final brightness of the pulp remained unchanged. Pretreatment with Cartazyme HS-10 also reduced the chlorine consumption by 31% in C stage, and the brightness ceiling was increased by 2.5 points (Table 6). At constant chemical dose, the final brightness of pulp was increased by 4.9, 3 and 2.1 points with Cartazyme HS-10, Novozyme 473 and VAI xylanase, respectively, in a CEH sequence. Bajpai et al. [61] also reported improved brightness of bamboo kraft pulp using xylanase in a conventional bleaching sequence, C_DEHD. It was possible to increase the final brightness of pulp by about 1.6 points with a final brightness of 88.8% PV versus 87.2% PV in the control. Alternatively, the enzyme could be used to decrease the active chlorine use in the first stage of the bleaching by 20% or decrease chlorine dioxide in the last stage of brightening by 4 kg t^{-1} in the C_DEHD sequence.

Table 6. Effect of different xylanase treatments on conventional bleaching of eucalyptus pulp

Treatment	CEHH	Novozyme-473 XCEHH	VAI Xylanase XCEHH	Cartazyme HS-10 XCEHH
Kappa No.	25.8 (control)	24.0 (after enzymatic pretreatment)	23.1 (after enzymatic pretreatment)	21.6 (after enzymatic pretreatment)
Chlorination				
Cl_2 added (%)	4.50	3.20	3.10	3.10
Cl_2 consumed (%)	4.48	3.08	3.08	3.08
Final pH	1.85	2.04	1.98	1.97
Extraction				
NaOH added (%)	0.8	0.8	0.8	0.8
Final pH	9.0	10.1	9.8	10.3
K. No. of pulp	5.3	5.7	5.4	4.7
Hypo-I				
Cl_2 added (%)	2.50	2.50	2.50	2.50
NaOH added (%)	0.50	0.50	0.50	0.50
Cl_2 consumed (%)	2.49	2.49	2.49	2.38
Final pH	7.7	7.4	6.9	7.4
Hypo-II				
Cl_2 added (%)	0.75	0.75	0.75	0.75
NaOH added (%)	0.40	0.40	0.40	0.40
Cl_2 consumed (%)	0.44	0.40	0.41	0.40
Final pH	10.5	10.6	10.5	10.6
Cl_2 added (%)	7.75	6.35	6.35	6.35
Cl_2 consumed (%)	7.41	5.97	5.98	5.86
Brightness (% ISO)	80.2	79.7	80.5	82.7
Viscosity (mPa s)	4.90	5.24	5.24	4.67
Yield (%)	92.00	91.71	91.71	90.00

% chemicals are based on oven-dried pulp; Cl_2 refers to active chlorine.
Conditions: X(Novozyme-473), 500 EXU kg^{-1}, pH 8.0, 40 °C, 60 min, 10% pulp; X(VAI Xylanase), 5000 XU kg^{-1}, pH 6–6.2, 60 °C, 120 min, 10% pulp; X(Cartazyme HS-10), 2000 XYU kg^{-1}, pH 4–4.5, 50 °C, 180 min, 10% pulp; C. pulp 3%, 30 °C, 45 min; E, pulp 10%, 55 °C, 90 min; H-I, pulp 10%, 40 °C, 150 min; H-II, pulp 10%, 40 °C, 60 min.
Based on data from Ref. 60

Allison et al. [62] examined the effectiveness of a new thermophilic enzyme, Cartazyme HT, which retains its activity at relatively high pH and temperature conditions. Experiments were performed to assess the effects of important pretreatment conditions on conventional bleaching, D/CED, of kraft and kraft oxygen pulp from radiata pine. The enzyme pretreatment was fairly robust to changes in pretreatment conditions. Bleaching improvements were higher than 20% at pretreatment temperatures between 60 and 80 °C and at pH 7–9. Kraft and kraft oxygen pulps achieved similar levels of improved bleaching after enzyme pretreatment. However, pretreatment removed nearly twice as many carbohydrates from kraft oxygen pulp as from kraft pulp. Allison et al. [63] also assessed the effects of pretreatment with Cartazyme HT on bleaching with ozone, oxygen and chlorine dioxide. Overall enzyme treatment of kraft oxygen pulp prior to DED and ZED bleaching was fairly robust to changes in pretreatment conditions. Improvements of 19 to 24% were observed after enzyme pretreatment at 70 °C and pH 8.0. Enzyme charge was the most consistent pretreatment variable to significantly affect DED and ZED bleaching, with increased charge improving subsequent oxygen delignification. The kappa number of pulp delignified with oxygen was reduced by only 4% when enzyme pretreatment was employed. Final bleaching was also unaffected by the initial enzyme pretreatment, showing that it may only be effective prior to acidic delignification treatments such as chlorination or ozonation.

Eriksson and Yang [64] and Yang et al. [65] also reported that pretreatment with xylanases improved the bleachability of both hardwood and softwood kraft pulps. Hardwood kraft pulps bleached in OXDP sequence reached a brightness of 87.7% ISO whereas without enzyme treatment, a brightness of only 85.7% was obtained. Softwood pulp, bleached in OXPDP sequence, attained a brightness of 88.6% whereas without using enzyme stage, a brightness of only 84.5% was obtained. Pekarovicova et al. [66] investigated the effect of water prehydrolysis of beechwood on bleaching of prehydrolyzed kraft pulp with xylanases. It was found that the effectiveness of enzyme bleaching increased with severity of water prehydrolysis, which is probably caused by increased accessibility and diminished xylan resorption into the fibre surface of prehydrolyzed pulps (Table 7).

Ragauskas et al. [67] examined the effect of xylanase pretreatment on bleaching efficiency for a variety of nonchlorine bleaching agents. It was found that xylanase pretreatment of softwood kraft pulps can enhance the bleaching efficiency of nonchlorine-based bleaching agents. Optimal biobleaching results were obtained with ozone which exhibited enhanced bleaching selectivity and brightness gains. Xylanase pretreatment also improved brightness and delignification of peracetic acid bleached pulp (Table 8).

Eriksson and Yang [68, 69] studied the combined effect of four nonchlorine bleaching stages i.e. oxygen, xylanase, ozone and hydrogen peroxide, on the bleaching of hardwood and softwood kraft pulps. When eucalyptus pulp was bleached in OXZP sequence using 0.8% ozone, a brightness of 90% ISO was readily obtained compared to a brightness of 84.7% for the control OZP pulp.

Table 7. Effect of water prehydrolysis on bleachability of xylanase-pretreated kraft pulps

Characteristic		Prehydrolysis time (min)						
		0	1	30	60	70	80	90
Weight loss (%)		–	3.3	11.8	18.1	20.0	20.7	22.1
Kappa Number	Before enzyme treatment	15.26	14.85	11.79	8.20	9.78	9.75	9.10
	After enzyme treatment	9.63	9.64	7.09	4.9	5.01	5.01	4.74
	After enzyme and alkaline treatment	8.25	7.85	5.15	2.98	2.89	2.86	2.76
Brightness (% ISO)	After enzyme treatment	30.94	29.71	30.87	34.17	32.75	33.32	33.45
	After enzyme treatment and alkaline extraction	35.65	35.55	40.54	45.91	45.95	43.85	45.94

Based on data from Ref. 66

Table 8. Effect of xylanase treatment on bleaching efficiency of peracetic acid and ozone

Bleaching treatment	Kappa number	Tappi Brightness (%)	Viscosity (mPa s)
Brownstock	22.3	24.0	14.2
Xylanase-treated brownstock	21.7	26.8	15.1
X(PAA-2% charge) E	7.2	51.4	7.6
(PAA-2% charge) E	8.8	49.4	7.8
X(PAA-3.9% charge) E	5.4	51.4	7.4
(PAA-3.9% charge) E	6.3	48.7	7.8
X(PAA-4.4% charge) E	5.8	51.2	8.9
(PAA-4.4% charge) E	6.2	49.8	8.5
Bleached with 1.2% ozone/pH 2.5			
Xylanase treated	9.9	—	—
Control	11.6	—	—
Followed by caustic extraction:			
Xylanase treated	6.3	10.3	48
Control	7.6	9.4	44
Bleached with 0.6% ozone/pH 2.5			
Xylanase treated	15.2	—	—
Control	15.8	—	—
Followed by caustic extraction:			
Xylanase treated	11.2	12.1	38
Control	12.5	11.4	36
Bleached with 0.6% ozone/pH 5.0			
Xylanase treatred	15.0	—	—
Control	16.1	—	—
Followed by caustic extraction:			
Xylanase treated	12.2	12.2	36
Control	13.9	17.7	32

Extraction: 0.12M NaOH for 1 h at 70 °C
Based on data from Ref. 67

When pine kraft and pine RDH pulps were bleached in $OXZ_1E_PZ_2P$ and XE_PZP sequence, respectively, brightness values of 81 and 85.7% were obtained compared to 71.3 and 76.3% for the control pulps, respectively.

Sacon and Yang also studied the combined effect of a four stage chlorine-free bleaching using oxygen, xylanase treatment, ozone and hydrogen peroxide in short and long kraft pulp fibres [70]. When eucalyptus pulp was bleached in the sequence OXPZ, a brightness of 90.2% ISO was obtained.

Allison and Clark [71] have examined the effects of pretreatment with commercial hemicellulase enzyme, Pulpzyme HA, from Novo Nordisk on both ozone and D/CED bleaching sequences. For both bleaching sequences, it was possible to improve bleaching effectiveness by about 25%. Selection of pretreatment conditions was very important, particularly for the D/CED sequence because of the cellulase activity in the preparation. Optimal pretreatment conditions for ozone bleaching were markedly different from those for chlorine-based bleaching. Bleaching selectivity was unaffected by the pretreatment process.

Honshu Paper Co. of Japan investigated the use of xylanase enzymes for reducing or eliminating the use of chlorine in subsequent bleaching [18]. Bleaching trials were carried out either by using the D/CEDHD sequence or without the use of molecular chlorine by using the DE_oDHD sequence. It was found that pretreatment with xylanase led to a reduction of 20–25% in the amount of chlorine or chlorine dioxide used in the first stage. Ledoux et al [72] described the use of bacterial xylanases in combination with TCF or ECF bleaching sequences to produce acceptable pulps with 88% ISO brightness [72].

In European Patent No. 373, 107, a process is described in which pulp can be bleached by treatment with an enzymatic system containing hemicellulase of *Aureobasidium pullulans* [73]. The hardwood kraft pulp suspension bleached with 100 units of endoxylanase (from *A. pullulans*) showed a kappa number of 7.8 compared with 12.3 for the unbleached kraft pulp.

In European Patent No. 0395792 [74], a process is described by Enso Gutzeit OY in which oxygen is used in the first oxidation stage of the bleaching process. The pulp is then treated with hemicellulase enzyme and washed. The subsequent oxidation stage is carried out using bleaching chemicals containing chlorine, e.g. chlorine gas and/or chlorine dioxide. The amount of chlorine in the bleaching process can be substantially reduced by subjecting the pulp to preliminary oxygen bleaching and enzyme treatment and replacing the part of the chlorine gas conventionally used in bleaching with chlorine dioxide. Furthermore, in the bleaching procedure, the liquid obtained from washing stages after oxygen and enzyme treatment can be treated in a soda recovery boiler. Thus, the use of enzyme enables the amounts of toxic compounds in the spent bleach liquor to be reduced while simultaneously lowering the COD (Table 9).

In another European Patent No. 0383999 [75], the same paper company has described another process for bleaching of pulp. The essential features are that the chemicals used in the oxidation stage have a chlorine dioxide content of at least 50%, that the pulp is subjected to enzyme treatment either in conjunction

Table 9. Effect of hemicellulase treatment on chlorine/chlorine dioxide bleaching

	OXOEDED	OXOED/C(80/20) ED	OXEDED	C/D(90/10) EDED	OC/D(80/20) EDED
Consumption (kg t^{-1} pulp)					
Active chlorine	87	62	95	100	73
Chlorine gas	—	10	—	63	30
Chlorine dioxide	34	20	37	13	16
AOX (kg t^{-1} pulp)	1.6	1.0	1.8	4.0	2.5
Reduction in AOX (%)	60	75	55	0	38
COD (kg t^{-1} pulp)	35	35	35	71	40
Reduction in COD (%)	51	51	51	0	44

Based on data from Ref. 74

Table 10. Effect of hemicellulase treatment along with chlorine/chlorine dioxide bleaching on effluent characteristics

Bleaching sequence	COD (kg t^{-1} pulp)	AOX (kg t^{-1} pulp)
No enzyme treatment, Cl$_2$ bleaching	58.1	2.6
No enzyme treatment, bleaching with mixture of 90% ClO$_2$–10% Cl$_2$	55.0	1.0
Enzyme treatment, bleachig with mixture of 90% ClO$_2$–10% Cl$_2$	40.0	0.6
Enzyme treatment, washing and bleaching with mixture of 90% ClO$_2$–10% Cl$_2$	40.0	0.6

Based on data from Ref. 75

with or before the oxidation, and that after the oxidation and enzyme treatment, the pulp is treated without alkali. The use of enzyme reduced the amount of chlorophenols and other forms of organically bound chlorine in the spent bleach liquor and COD (Table 10).

The patent rights of Enso Gutzeit's enzymatic bleach boosting technology are to be taken over by the Finnish companies Alko Biotechnology and J.P. International and the Danish company Novo Nordisk. The xylanase-based enzymes from Novo Nordisk and Alko will have their efficiency tested in Enso's kraft pulping process [76].

In U.S. Patent No. B,179,021, a process for bleaching pulp is described which comprises of an oxygen bleaching treatment and treatment with an essentially cellulase-free xylanase [77]. The pulp is then subjected to conventional bleaching, C/DED. The process provides a delignified and bleached pulp using lower amounts of chlorine-containing compounds and affords the opportunity to eliminate the use of elemental chlorine, thereby reducing the pollution load.

Also, a greater extent of delignification is achieved while retaining acceptable pulp strength properties.

Johnson examined new processes for bleaching including pretreatment with xylanase enzymes, 100% substitution of chlorine with chlorine dioxide, reinforced extraction, oxygen delignification, and use of ozone and peroxide as well as modified continuous cooking and modified batch cooking for producing ECF or TCF pulps [78].

Strunk et al. [79] showed that the application of enzyme technology in combination with the use of hydrogen peroxide-reinforced extraction stages on softwood kraft pulp allows a lower kappa factor in 100% chlorine dioxide bleaching sequences and thus, the added benefit of decreased AOX and colour in the effluent. Elm et al. [80] carried out similar studies with hardwood pulp. Results showed that bleaching with lower kappa factors in $DE_{OP}DE_PD$ or XDE_ODE_PD sequences proportionally lowered effluent colour and AOX. Effluent colours were lowered by 20% by using a 0.10 kappa factor instead of a 0.26 kappa factor. Similarly, AOX decreased by 44% to a total of 1.1 kg t^{-1}. The use of H_2O_2 and xylanase at 0.15 kappa factor resulted in pulp with the same or higher brightness, equal AOX discharge and significantly higher viscosity as compared to oxygen delignification. The bleach plant chemical costs for DE_ODE_PD or XDE_ODE_PD sequences were higher than ODE_ODED. However, if the capital costs for the changes are taken into account, peroxide and enzyme bleach boosting results in a saving of $6 to $7 per ton as compared to oxygen delignification.

Paloheima et al. [81] reported that the bleachability of softwood kraft pulp delignified with oxygen was improved when the pulp was pretreated with xylanase obtained from genetically engineered strains of *Trichoderma reesei* and then subjected to ECF bleaching. Buchert et al. [82] studied the role of xylanases and mannanases in the treatment of softwood pulp prior to bleaching. It was found that xylanases improved bleachability in both the delignification and brightening stages while mannanases acted by a different mechanism and were beneficial mainly to delignification. Improvements were seen in both stages when the two enzymes were used together.

Xylanase pretreatment has led to reductions in effluent adsorbable organic halogen (AOX) and dioxin concentrations due to reduced chlorine requirement to achieve a given brightness [83–85]. The level of AOX in effluents was significantly lower for xylanase pretreated pulps compared to conventionally bleached control pulps [52] (Table 11). When softwood kraft pulp pretreated with xylanase was bleached to 90% ISO using a C/DE_PDED sequence, the required kappa factor was reduced from 0.22 to 0.15 which was below the kappa factor of 0.18–0.19 required for the formation of chlorinated dioxins and furans [10]. The AOX concentration in a combined effluent stream was reduced by 33% compared to the control [83]. The effluent biochemical oxygen demand (BOD) doubled and there were increases in effluent chemical oxygen demand (COD) and total organic carbon (TOC). The BOD/COD ratio also increased indicating that the effluent was more amenable to biological degradation in

Table 11. Effect of xylanase treatment on the AOX content of the $(C + D)E_PDED$ effluents at 20% chlorine dixoxide substitution

Pulp	Chlorine charge on pulp (%)	AOX (kg t^{-1} pulp)	Brightness (% ISO)
Control	5.2	4.35	
	6.3	6.00	91.2
Xylanase	4.3	3.00	
	5.2	3.75	91.0
	5.7	4.50	

Based on data from Ref. 52

a secondary treatment plant. Effluent toxicity remained essentially the same. In the same study, xylanase pretreatment of a hardwood kraft pulp under the same conditions led to a reduction of 35–40% in chlorination charge. E_1 stage AOX was 24% less than in the control and the BOD/COD ratio was increased. Also, the organochlorine content of the pulp was reduced by 41% at a chlorine dioxide substitution level of 40% [83]. Bajpai et al. [59, 60] reported that the total organic chlorine (TOC1) content in extraction stage effluent was reduced by 30% when the pulp was first pretreated with xylanase and then subjected to CEH bleaching.

Enzymatically treated pulps show unchanged or improved strength properties [11, 32, 59, 60]. Also these pulps are easier to refine than the reference pulps. Improved viscosity of the pulp has been noted as a result of xylanase treatment [36, 52, 59–61, 65]. This is probably caused by the selective removal of xylan as found out by the pentosan values. Xylan with lower DP than cellulose can be expected to lower the average viscosity of kraft hemicellulose. However, the viscosity of the pulp was adversely affected when cellulase activity was present [59, 71, 86, 87]. Therefore, the presence of cellulase activity in the enzyme preparation is not desirable. In a few cases, lower mechanical strength has been obtained with xylanase-treated pulp, probably due to the presence of cellulase in the enzyme preparation [88].

7.2 Plant Scale Trial with Xylanases

Canfor's International mill in British Columbia, Canada, was able to cut the use of chemicals and production of effluents by employing an enzyme, Irgazyme 10S, manufactured by Genencor International. Total chlorine dioxide consumption was reduced by 15.6% or 6.4 kg t^{-1} of pulp for the entire period. The active chlorine multiple was lowered by 28.3% from the control period average of 0.265 to the trial average of 0.19. The pulp quality of the bleached enzymatically treated pulp in terms of cleanliness was equal to or better than normal fully bleached pulp [89].

In Tasman Pulp and Paper Company Ltd., New Zealand, which produces about 500 t d^{-1} of bleached pulp from *Pinus radiata* using a D/CE$_0$DED sequence, a mill trial was performed with Genencor's second generation product – Irgazyme 40S-4X – which showed improved performance at high pH and temperature. The total chlorine dioxide requirement was found to be reduced by 20% with improvements in AOX level and pulp qualities like brightness [90, 91].

Morrum Pulp Mill in Sweden used Novo Nordisk's second generation product, Pulpzyme HB, to reduce total chlorine consumption. Instead of 48 kg of chlorine dioxide per ton of pulp, only 37 kg were needed to obtain the same brightness. Also there was a corresponding reduction in AOX level [27, 29].

In another mill trial with Pulpzyme HB and softwood pulp, the total active chlorine demand was found to be reduced by 12%. A reduction of 23–33% in AOX levels was achieved by enzyme treatment when periods before and after the trial were compared [27, 29].

Bukoza Pulp Mill in Vranov CSFR, which produces 220 t d^{-1} of bleached kraft pulp from beech using a C/DEDE$_H$D sequence, verified the effectiveness of Voest Alpine xylanase. The consumption of chlorine was reduced by 30% and hypochlorite by 38% using the enzyme [92]. The consumption of other bleaching chemicals as well as final brightness, physical properties and yield remained unchanged.

Metsa-Sellu Mill in Aanekoksi (Finland) used 35 tons of enzyme, Albazyme 10, in a 4 week long trial for the production of 35 000 tons of fully bleached pulps derived from softwoods and hardwoods with a total active chlorine saving of 12%. The output was used for paper production by the Metsa-Serla group companies [93].

Enso Gutzeit OY, a Finnish company, ran enzymatic bleaching tests on a scale of 1000 m^3. Chlorine consumption was found to be reduced by 25–30% [94]. Another plant scale trial using an 800-ton-run resulted in a 12% reduction in chemical requirements in a softwood kraft mill with a D/CEDED bleaching sequence [95].

The Donohue St. Felicien mill in Quebec, which produces 900 t d^{-1} of ECF softwood kraft pulp, used xylanase treatment to reduce the amount of bleaching chemicals [96]. It was found that use of xylanase reduced the kappa factor from 0.165 to 0.12–0.13 while subsequent bleaching still produced 91% ISO brightness and good strength properties. About 0.5% less chlorine dioxide was required and AOX was reduced to below 0.3 kg t^{-1}. When xylanase was combined with peroxide in the first extraction stage, the organic halogen content of the pulp was reduced from 165 ppm to 100–110 ppm.

In a 3-day-mill trial with Cartazyme HS-10, in which 1700 tons of mixed hardwood kraft pulp were processed, about 30% savings in active chlorine occurred in D/C stage when the enzyme was introduced in the $D_{75}C_{25}E_0D_1E_PD_2$ sequence. The properties of bleached pulp at different degrees of refining showed improvements when compared with normal operating conditions [97].

The capability of the enzyme to reduce the consumption of bleaching chemicals makes it possible to consider significant modifications in the bleaching sequence. It is possible to completely exclude the first chlorination stage in $C/DE_{OP}DE_PD$ and replace it with an enzyme state (X) to become $XE_{OP}DE_PD$. This has been verified in a mill scale trial [11, 98, 99]. The advantage of this modified sequence is that the filtrates from the E_{OP} stage can be recirculated to the recovery system without risk of chloride-initiated corrosion. This will, therefore, contribute to closing the water circulation system of the pulp mill and minimize the discharge of effluents. The following sequence options were tested on hardwood kraft pulp with an incoming kappa number of approximately 14 and a viscosity of $850-950$ dm^3 kg^{-1}.

1. $XE_{OP}DE_PD$ (Enzyme/no chlorine)
2. $E_{OP}DE_PD$ (No enzyme/no chlorine)

The advantage of options 1 and 2 is that the filtrate from the prebleaching can be kept separate from the filtrates of the final bleaching and will, therefore, not contain components that could not be evaporated and burned in the recovery boiler. The effluent load will, in these cases, be lower than for the conventional sequences, even if the final bleaching in options 1 and 2 will require slightly higher chlorine dioxide charges. The mill had previously tested the $E_{OP}DE_PD$ sequence without enzyme and could not achieve higher brightness values than 83% ISO. With enzyme, a brightness level of 88% was achieved (Table 12).

Cost reduction is of great importance, especially in TCF bleaching. In a mill scale trial with Cartazyme HS-10, it was found that peroxide could be saved. Compared to the QPP sequence, the XQPP sequence could save $5-10$ kg H_2O_2 t^{-1} of pulp [11].

In 1992, in a joint venture between Korsnas AB and the Dutch company Gist brocades, Korsnas started a full scale test on TCF bleaching using the enzyme Korsnas T6 xylanase. The enzyme, which was completely free of cellulase side activity, was isolated from *Bacillus stearothermophilius*. It worked well at high pH and temperature and showed a good storage stability. With this enzyme, it was possible to produce TCF pulp of 78% ISO brightness [20].

Table 12. Elemental chlorine-free bleaching of hardwood pulp with xylanase (Cartazyme HS-10)[a]

Parameter	$E_{OP}DE_PD$	$XE_{OP}DE_PD$
Pulp production (t d^{-1})	380	380
Chemical addition (kg t^{-1})		
Chlorine dioxide	12.7	13.8
Hydrogen peroxide	12.7	12.5
Final brightness (% ISO)	81.5	88.2
Unbleached kappa number	13.8	14
E_{OP} viscosity (dm^3 kg^{-1})	799	809

[a]Results of plant scale trial
Based on data from Ref. 98

Recently, Aanekoski mill in Finland used an enzyme with oxygen delignification and hydrogen peroxide bleaching to produce over 50 000 tons of totally chlorine-free pulp [100]. A patent has been filed on this process.

A new pilot plant has come online at the University of Georgia campus in Athens, Georgia, to bridge the transition from laboratory to commercial production of the enzyme/ozone bleaching process. The process, trade marked Enzone with patent pending, features an oxygen-xylanase-ozone-hydrogen peroxide sequence for hardwood pulp and an additional alkaline extraction stage between ozone and hydrogen peroxide stages for softwood pulps. The main advantage of this process is the xylanase step which increases brightness by 3 to 8 points when compared with pulps bleached without it [101].

A significant number of European, North American and Japanese mills are presently bleaching full time with enzymes. Crestbrook Forest Industries, British Columbia, Canada, use the enzyme Albazyme 10 to remove the bottleneck in chlorine dioxide generation and to increase the production of ECF pulp. Munksjo AB Sweden combine the enzyme with the Lignox process to produce TCF pulps of 73–75% ISO brightness, while Metsa-Sellu OY, Finland, use it to reduce adsorbable organic halogen (AOX) emissions from its kraft pulp mill. Albazyme 10 is presently available from Cultor, Finland; Biopulp International, France; and Genencor International, U.S.A. [102, 103]. Metsa-Botnia's plant in Kasko now produces chlorine-free spruce pulp by using several oxygen delignification steps and enzymatic bleaching. The pulp is used in the production of supercalendered light-weighted coated paper and tissue paper. Metsa-Botnia hope to develop the process into a totally closed system and to be able to bleach birch pulp in similar ways [104].

As the TCF pulp market grows, Canadian mills, which are the world's largest exporters of market pulp, have started to investigate chlorine-free bleaching which involves the use of xylanase enzymes. European paper makers are now requesting TCF pulps or bleaching pulps with extremely low AOX and/or TOX level in their effluents [105].

Many Finnish companies-Enso Gutseit OY, Kimi Kymmene, Metsa–Sellu OY, Sunila OY, Veitsiluoto OY and United Paper Mills — are conducting research to develop chlorine-free bleaching which involves the use of enzymes, oxygen, peroxide and ozone [106].

8 Benefits from Xylanase Treatment

Xylanase pretreatment of pulps prior to bleach plant reduces bleach chemical requirements and permits higher brightness to be reached. The reduction in chemical charges can translate into significant cost savings when high levels of chlorine dioxide and hydrogen peroxide are being used. A reduction in the use of chlorine chemicals clearly reduces the formation and release of chlorinated organic compounds in the effluents and the pulps themselves. The ability of xylanases to activate pulps and increase the effectiveness of bleaching chemicals, may allow new bleaching technologies to become more effective. This means that for expensive chlorine-free alternatives such as ozone and hydrogen peroxide, xylanase pretreatment may eventually permit them to become cost effective.

Traditional bleaching technologies also stand to benefit from xylanase treatments. Xylanases are easily applied and require essentially no capital expenditure. Because chlorine dioxide charges can be reduced, xylanase may help eliminate the need for increased chlorine dioxide generation capacity. Similarly, the installation of expensive oxygen delignification facilities may be avoided. The benefit of a xylanase bleach-boosting stage can also be taken to shift the degree of substitution towards higher chlorine dioxide levels while maintaining the total dosage of active chlorine. Use of high chlorine dioxide substitution dramatically reduces the formation of TOCl. Xylanases may also be the ticket to success for a chlorine-free bleaching alternative.

Xylanase technology has been catapulated from biotechnology labs to pulp mills in just a few short years. The main driving factors have been the economic and environmental advantages the enzyme brings to the bleach plant. Such intense demand for the enzyme has pushed enzyme producers to develop an entirely new industry in a remarkably short time. The increasing competition among manufacturers will continue to improve products and reduce the price.

9 Future Developments

Enzyme technology for pulp bleaching will continue to develop at a rapid pace. Already xylanase products covering various temperature and pH ranges are making their way into the market. This is guided by the fact that the optimal conditions for an enzyme depend on both parameters, which means that a high temperature lowers the optimum pH of the enzyme and a high pH equally lowers the optimum temperature. The results will be that an enzyme with both high pH and temperature optima is needed to match the conditions occurring in the pulping process.

Enzymes for pulp bleaching can not yet be considered as commodity chemicals as there can be significant differences between commercially available products regarding activity, contamination with unwanted enzymes (e.g. cellulases in a predominantly xylanase product), stability during storage and at the temperature and pH of the application, consistency and quality of product supply etc. Therefore, there is a considerable room for development in the research, manufacture, downstream processing and application knowledge of enzymes.

The bleaching process will be affected by the move to reduce the impact of the entire pulping process on the environment. How quickly will chlorine chemicals be abandoned and oxygen-based chemicals (oxygen, hydrogen peroxide and ozone) be substituted? Who will pay for the increased investment or higher costs that result? How successful will enzymes be at providing a long term approach to reducing costs and investment? There are research and development efforts focussed on delignification before bleaching; notable is the progress with extended cooking and oxygen delignification which makes ozone bleaching more attractive, and this may encourage the growth of the TCF pulp market.

Those looking strategically at the future see the need to look at the ecological balance of the entire mill, predicting that by the year 2000, there will be a market for a new product, total effluent free pulp (TEF).

10 Conclusions

In the rapidly changing field of pulp bleaching, many efforts have been made to adjust the process to meet economical and environmental realities. Many of the new technologies have not yet proved feasible, especially as they often include major investment and negative effects on pulp properties. In this situation, enzyme bleaching has turned out to be a good alternative as it

- adapts easily to both ECF and TCF bleaching,
- offers cost savings,
- requires minimal investment,
- maintains or improves pulp quality,
- is applicable to a variety of chemical pulps.

Acknowledgements. We thank various investigators for sharing their papers awaiting publication as well as unpublished results; this has allowed us to extend coverage somewhat beyond late 1994. We also thank Mr. S.S. Gill for typing the manuscript.

11 References

1. Rydholm AA (1965) Chemical pulping. In: Pulping Processes Wiley Interscience, New York, p 575
2. Sanyer M and Chidester CH (1963) Manufacture of wood pulp. In: Chemistry of Wood, Browing BL (ed), Wiley Interscience, New York, p 441
3. Gierer J (1970) The reaction of lignin during pulping: A description and comparison of conventional pulping processes. Sven Papperstid, 73: 571
4. Gierer J (1981) Chemical aspects of delignification. In: Proceedings of Ekman Days International Symposium on Wood and Pulping Chemistry, Vol. 2, Swedish Forest Products Laboratory, Stockholm, Sweden, p 12

5. Ljunggren S (1980) The significance of arylether cleavage in kraft delignification of softwood. Svensk Papperstidn, 83: 363

6. Marton J (1991) Reactions of lignin in alkaline pulping. In: Lignins: Sarkanen KV, Ludwig CH (eds) Occurrence, Formation, Structure and Reaction. Wiley Interscience, New York, p 639

7. Yamasaki T, Hosoya S, Chen CL, Gratzl JS and Chang HM (1981) Characterization of residual lignin in kraft pulp. In: Proceedings of Ekman Days International Symposium on Wood and Pulping Chemistry, Vol. 2, Swedish Forest Products Laboratory, Stockholm, Sweden, p 34

8. Falkehag I, Marton J, Adler E (1956) Adv Chem Ser 59: 75

9. Trubacek I and Wiley A (1979) Bleaching and pollution. In: Bleaching of Pulp, Singh RP (ed.) Tappi Press, Atlanta, GA, p 423

10. Berry RM, Fleming BI, Voss RH, Luthe CE and Wrist PE (1989) Towards preventing formation of dioxins during chemical pulp bleaching, Pulp Paper Canada, 90(8): T279

11. Dunlop N and Gronberg V (1994) Recent developments in the application of xylanase enzymes in ECF and TCF bleaching. 80th Annual Meeting Technical Section, Montreal, Canada, February 1–2, Preprints A, p A191

12. Teras T (1992) TCF Marketing aspects, Asia Pacific Pulp Paper, August

13. Thayer AM (1993) Product report, Paper chemicals, Chemical and Engineering News, 28–41, November

14. Anonymous (1991) Enzymes are breaking into paper. Pulp and Paper International, 33(9): 81

15. Koponen R (1991) Enzyme system prove their potential. Pulp and Paper International, 33(11): 20

16. Eriksson KE (1992) Biotechnology will play key role in future nonchlorine bleaching. Pulp and Paper, 66(2): 149

17. Liebergott N, Van Lierop B, Fleming BI (1992) Lowering AOX levels in the bleach plant. In: Proceedings of Tappi Environmental Conference, Richmond, U.S.A. April 12–15, Book 3, Tappi Press, Atlanta, p 1065

18. Wakai M, Kai K and Ohera Y (1992) Less chlorine and nonchlorine bleaching with enzymatic treatment. In: Proceedings of Pan Pacific Pulp and Paper Technology Conference, Tokyo, Japan, Sept. 8–10, Part A, p 47

19. Lavielle P (1993) Xylanase prebleaching, Asia Pacific Papermaker, 3(5): 29

20. Sandstrom AS (1993) Total chlorine free bleaching using enzymes. Svensk Papperstidn, 7: 40

21 Lavielle P (1993) Xylanase prebleaching technology - an innovative answer to chlorine less and chlorine free bleaching of kraft pulps. In: Proceedings of EUCEPA International Environmental Symposium-Pulp and Paper Technology for a Cleaner World, Paris, France, April 27–29, Vol. I, p 151

22. Gronberg V (1993) Chlorine free bleaching of chemical pulp by the use of enzymes, Tissue World, 93, Nice, France, March 2–4, p 10

23. Gronberg V (1993) Chlorine free bleaching of chemical pulp by the use of enzymes. Paper South Africa, 13(6): 12

24. Gronberg V, Farell RL and Skerker PS (1993) Biobleaching. In: Proceedings of XXV EUCEPA Conference Pulp and Paper 2000, Vienna, Australia, October 4–8, p 167

25. Freiermuth B, Koljonen M and Werthemann P (1993) Enzymatic prebleaching of kraft pulp: Innovative technology to decrease the demand for bleaching chemicals and reducing AOX discharges, Progress 93 – Needs and Possibilities of Paper Industry Development in the Countries Changing their Economic System, Lodz, Poland, Sept. 27–30, Vol. I, p 218

26. Bajpai P and Bajpai PK (1992) Biobleaching of kraft pulp. Process Biochemistry, 27, 319

27. Holm HC, (1992) Recent progress and mill scale experiences with enzymatic bleach boosting of kraft pulp. In: Proceedings of Pan Pacific Pulp and Paper Technology Conference, Tokyo, Japan, September 8–10, Part A, p 53

28. Grant R (1993) R&D optimizes enzyme applications. Pulp and Paper International, 35(9): 56

29. Munk N (1993) Bleach boosting with xylanases: recent research results. In: Proceedings of 47th Appita Annual General Conference, Rotorua, New Zealand, April 19–23, Vol. 1, p 257

30. Daneault C, Leduc C and Valade JL (1994) The use of xylanases in kraft pulp bleaching. Tappi Journal 27(6): 125

31. Paice MG and Jurasek L (1984) Removing hemicellulose from pulps by specific enzyme hydrolysis. Journal of Wood Chemistry and Technology, 4(2): 187

32. Viikari L, Ranua M, Kantelinen A, Sundquist J and Linko M (1986) Bleaching with enzymes. In: Proceedings of Third International Conference on Biotechnology in Pulp and Paper Industry, Stockholm, Sweden, p 67

33. Viikari L, Ranua M, Kantelinen A, Linko M, Sundquist J (1987) Application of enzymes in bleaching. In: Proceedings of 4th International Symposium on Wood & Pulping Chemistry, Paris, April 27–30, Vol. 1, p. 151

34. Kantelinen A, Rättö M, Sundquist J, Ranua M, Viikari L and Linko M (1988) Hemicellulases and their potential role in bleaching. In: Proceedings of Tappi International Pulp Bleaching Conference, Orlando, June 5–9, p 1

35. Jurasek L, Paice M (1992) Saving bleaching chemicals and minimizing pollution with xylanase. In: Proceedings of the International Symposium on Pollution Prevention in the Manufacture of Pulp & Paper-Opportunities and Barriers, Washington, DC., 1992

36. Paice MG, Bernier R, Jr. and Jurasek L (1988) Viscosity enhancing bleaching of hardwood kraft pulp with xylanase from a cloned gene. Biotechnology Bioengineering, 32: 235

37. Viikari L, Kantelinen A, Rättö M and Sundquist J (1991) Enzymes in pulp and paper processing. In: Enzymes in Biomass Conversion, Leatham GF and Himmel ME (eds) ACS Symposium Series 460, American chemical society. Washington, DC., p 12

38. Viikari L, Sundquist J and Kettunen J (1991) Xylanase enzymes promote pulp bleaching, Paper and Timber, 384

39. Kantelinen A, Hortling B, Sundquist J, Linko M and Viikari L (1993) Proposed mechanism of the enzymatic bleaching of kraft pulp with xylanases. Holzforschung, 47: 318

40. Paice MG, Gurnagul N, Page DH and Jurasek L (1992) Mechanism of hemicellulose directed prebleaching of kraft pulp. Enzyme and Microbial Technology, 14: 272

41. Clayton DW and Stone JE (1967) The redeposition of hemicellulose during pulping. Part 1: The use of a tritium-labelled xylan. Pulp and Paper Magazine of Canada, 64: T 459

42. Pederson LS, Kihlgren P Nissen AM, Munk N, Holm HC and Choma PP (1992) Enzymatic bleach boosting of kraft pulp: laboratory and mill scale experience. In: Proceedings of Tappi Pulping Conference, Boston, M.A., U.S.A., November 1–5, Book 1, Tappi Press, Atlanta, p 31

43. Barnard F, Comtat J, Joseleau JP, Mora F and Ruel K (1986) Interest in the enzymatic hydrolysis of xylanase for modifying the structure of pulp fibres. In: Proceedings of Third International Conference on Biotechnology in Pulp and Paper Industry, Stockholm, Sweden, June 16–19, p 70

44. Tan LUL, Yu EKC, Bouis-Seize GW and Saddler JN (1987) Inexpensive, rapid procedure for bulk purification of cellulase-free β-1,4-D-xylanase of high specific activity. Biotechnology Bioengineering, 15: 96

45. Tolan JS (1992) The use of enzymes to enhance pulp bleaching. In: Proceedings of Tappi Pulping Conference, Boston, MA, USA, November 1–5, Book 1, Tappi Press, Atlanta, p 13

46. Tolan JS (1992) Mill implementation of enzyme treatment to enhance bleaching. In: Proceedings of 78th CPPA Annual Meeting, Montreal, Canada, January 28–29, p A163

47. Tolan JS and Canovas RV (1992) The use of enzymes to decrease the Cl_2 requirements in pulp bleaching, Pulp and Paper Canada, 93(5): 39

48. Du Manoir JR, Hamilton J, Senior DJ, Bernier RL, Grant JE, Moser LE, Dubelsten P (1991) Biobleaching of kraft pulps with cellulase free xylanase. In: Proceedings of International Pulp Bleaching Conference, Stockholm, Sweden, June 11–14, Vol 2, p 123

49. Jurasek L and Paice MG (1958) Biological treatment of pulps. Biomass, 15: 103

50. Bernier R Jr, Driguez H and Desrochers M (1983) Molecular cloning of a Bacillus subtilis xylanase gene in Escherichia coli. Gene, 26: 59

51. Senior DJ and Hamilton J (1992) Bleaching with xylanases brings biotechnology to reality, Pulp and Paper, 66(9): 111

52. Senior DJ, Hamilton J, Bernier RL and du Manoir JR (1992) Reduction in chlorine use during bleaching of kraft pulp following xylanase treatment, Tappi Journal, 75(11): 125

53. Senior DJ, Hamilton J (1993) Xylanase treatment for the bleaching of softwood kraft pulps: The effect of chlorine dioxide substitution. Tappi Journal, 76(8): 200

54. Senior DJ, Hamilton J (1992) Xylanase treatment for the bleaching of softwood kraft pulps: The effect of chlorine dioxide substitution. In: Proceedings of Tappi Pulping Conference, Boston, USA, November 1–5, Book 1, Tappi Press, Atlanta, p 19

55. Skerker PS and Farell RL (1991) Chlorine free bleaching with Cartazyme HS treatment. In: Proceedings of International Pulp Bleaching Conference, Stockholm, Sweden, June 11–14, Vol. 2, p 93

56. Clark TA, Steward D, Bruce ME, Mcdonald AG, Singh AP and Senior DJ (1991) Improved bleachability of radiata pine kraft pulps following treatment with hemicellulolytic enzymes. Appita, 44(6): 389

57. Pederson SL, Nissen AM, Elm DD and Choma PP (1991) Bleach boosting of kraft pulp using

alkaline hemicellulases. In: Proceedings of International Pulp Bleaching Conference, Stockholm, Sweden June 11–14, 1991, Vol. 2, p 107
58. Gibson K (1992) Enzymatic bleach boosting, 25th Annual Pulp and Paper Meeting, Saupaulo, Brazil, November 23–27, p 47
59. Bajpai P, Bhardwaj NK, Maheshwari S and Bajpai PK (1993) Use of xylanase in bleaching of eucalypt kraft pulp. Appita, 46(4): 274
60. Bajpai P, Bhardwaj NK, Bajpai PK and Jauhari MB (1994) The impact of xylanases in bleaching of eucalyptus kraft pulp, Journal of Biotechnology, 36(1): 1
61. Bajpai P and Bajpai PK (1995) Application of xylanases in bleaching of bamboo kraft pulp, Tappi Journal – In press
62. Allison RW, Clark TA and Wrathall SH (1993) Pretreatment of radiata pine kraft pulp with a thermophillic enzyme Part I. Effect on conventional bleaching, Appita, 46(4): 269
63. Allison RW, Clark TA and Wrathall SH (1993) Pretreatment of radiata pine kraft pulp with a thermophillic enzyme Part II. Effect on oxygen, ozone and chlorine dioxide bleaching, Appita 46(5): 349
64. Eriksson KE and Yang JL (1992) Use of enzymes as one stage in pulp bleaching, In: Proceedings of Tappi Environmental Conference, Richmond, USA., April 12–15, 1992, Book 2, Tappi Press, Atlanta, p 411
65. Yang JL, Lou G and Eriksson KE (1992) The impact of xylanases on bleaching of kraft pulps, Tappi Journal 75(12): 95
66. Pekarovicova A, Rybarikova D, Kasik M and Fiserova M (1993) Prebleaching of kraft pulp by xylanases, The effect of water prehydrolysis, Tappi Journal, 76(11): 127
67. Ragauskas AJ, Pou KM and Cesternino AJ (1994) Effect of xylanase pretreatment procedures on non-chlorine bleaching. Enzyme Microbial Technology, 16(6): 492
68. Eriksson KE and Yang JL (1993) Enzyme augmentation of kraft pulp bleaching with oxygen based chemicals. In: Proceedings of Tappi Environmental Conference, Boston, M.A., USA, March 28–31, Book 2, Tappi Press, Atlanta, p. 627
69. Eriksson KE and Yang JL (1994) Bleaching of soft wood pulp with the enzyme process, Tappi Journal, 77(3): 243
70. Sacon UM and Yang JL (1994) Bleaching eucalyptus pulp with sequences containing oxygen, xylanase, ozone and peroxide, Papel. 2, 20
71. Allison RW, Clark TA (1992) Effect of enzyme treatment on ozone bleaching. In Proceedings of Pan-Pacific Pulp and Paper Technology Conference, Tokyo, Japan September 8–10, Part A, p 15
72. Ledoux P (1993) Use of bacterial xylanases in chlorine free bleaching sequences. In: Proceedings of Tappi Pulping Conference, Atlanta, GA, USA, November 1–3, Book 3, Tappi Press, Atlanta, p 1057
73. Farrel RL (1990) Bleaching of pulps using Aureobasidium pullulans cultures. EP 373, 107
74. Enso-Gutzeit OY (1990) Procedure for the bleaching of pulp, EP 0395, 792
75. Enso-Gutzeit OY (1990) Procedure for the bleaching of pulp, EP 0383, 999
76. Anonymous (1992) Enso bleach boost handover. Paper 27(3): 16
77. GIL Inc (1993) North York, Canada. Pulp bleaching process comprising oxygen delignification and xylanase treatment, US 5,179,021
78. Johnson AP (1993) Fitting together the ECF-TCF Jigsaw. In: Proceedings of 47th Appita Annual General Conference, Rotorua, New Zealand, April 19–23, Vol. 1, p 441
79. Strunk WG, Klein RJ, Elm DD, Choma PP, Sundaram VSM (1992) Enzyme boosting and peroxide reinforcement in 100% ClO_2 sequences - a low capital alternative to oxygen delignification. In: Proceedings of Tappi Pulping Conference, Boston, U.S.A., November 1–5, Book 1, Tappi Press, Atlanta, p 117
80. Elm DD, Choma PP, Strunk WG, Klein RJ and Sundaram VSM (1993) Enzyme bleach boosting and peroxide reinforcement of hardwood kraft in 100% chlorine dioxide bleaching sequences - a low capital alternative to oxygen delignification. 79th Annual Meeting Technical Section, Montreal, Canada, January 26–29, Preprints B, p B183
81. Paloheima M (1991) The use of xylanases from genetically engineered Trichoderma strains to improve the bleachability of kraft pulp. In: Proceedings of Seventh International Symposium on Wood and Pulping Chemistry, Beijing, China, May 25–28, Vol. 2, p 993
82. Buchert J (1993) The role of Trichoderma reesei, xylanase and mannanase in the treatment of softwood kraft pulp prior to bleaching, Holzforschung, 47(6): 473
83. Senior DJ and Hamilton J (1992) Use of xylanases to decrease the formation of AOX in kraft pulp bleaching. Journal of Pulp and Paper Science, 18(5), J165

84. Senior DJ and Hamilton J (1991) Use of xylanase to decrease the formation of AOX in kraft pulp bleaching. In: Proceedings of the Environmental Conference 1991 of the Technical Section, Canadian Pulp and Paper Association, Quebec, Canada, October 8–10, p 63

85. Vaheri M, Miiki K, Tokela V, Kitunen V and Salkinojasalonen M (1989) Bleaching of kraft pulp without the formation of dioxins, 9th International Symposium On Chlorinated Dioxins and Related Compounds, Toronto, Canada

86. Clark TA, Mcdonald AG, Senior DJ and Meyers PR (1990) Mannanase and xylanase treatment of softwood chemical pulps: Effects on pulp properties and bleachability. In Biotechnology in Pulp and Paper Manufacture Kirk TK and Chang HM (eds) Chap 14, Butterworth Heinmann, Raleigh NC, p 157

87. Puls J, Poutanen K and Lun J (1990) Enzymatic accessibility of hemicellulase in hardwood pulps to enzymes. In Biotechnology in Pulp and Paper Manufacture Kirk TK and Chang HM (eds) Chap 16, Butterworth Heinmann, Raleigh NC., p 183

88. Chauvet JM, Comtat J and Nue P (1987) Assistance in bleaching of never dried pulps by the use of xylanase, consequences on pulp properties. In: Proceedings of the 4th International Symposium on Wood and Pulping Chemistry, Paris, April 27–30, Vol. 2, p 325

89. Scott BP, Young F and Paice MG (1993) Mill scale enzyme treatment of a softwood kraft pulp prior to bleaching. Pulp and Paper Canada, 94(3): 57

90. Werthemann DP, Tanner D and Koljonen M (1993) Enzymatic pre-bleaching of Pinus radiata pulp, a technology to reduce AOX, 47th Appita Annual General Conference, Rotorua, New Zealand, April 19–23, Vol. 1, p 249

91. Werthemann D (1993) Pre-bleaching of Pinus radiata pulp using enzymes – Technology to reduce AOX, Japanese Journal of Paper Technology, 10: 15

92. Siiner M and Preselmayr W (1992) Chlorine is out, bring in the enzymes. Pulp Paper International, 34(9): 87

93. Grant R (1991) First mill scale trials get underway. Pulp Paper International, 33(6): 61

94. Trotter PC (1990) Biotechnology in the pulp and paper industry: a review, Tappi Journal, 73(4): 198

95. Onysko KA (1993) Biological bleaching of chemical pulps: A Review, Biotechnology Advances, 11: 179

96. Jean P, Hamilton J, Senior DJ (1992) Mill trial experiences with xylanase: AOX and chemical reductions. 80th Annual Meeting Technical Section, Montreal Canada, February 1–2, Preprints A, p A229

97. Perroloz JJ (1993) Cartazyme HS mill trial, Personal Communication, November, 1993.

98. Turner JC, Skerker PS, Burns BJ, Howard JC, Alonso MA and Andres JL (1992) Bleaching with enzymes instead of Cl_2. Tappi Journal, 75(12): 83

99. Skerker PS, Labbauf MM, Farell RL, Beerwan N and McCarthy P (1992) Practical bleaching using xylanases: Laboratory and mill experience with Cartazyme HS-10 in reduced and chlorine free bleach sequences. In: Proceedings of Tappi Pulping Conference, Boston MA, USA, November 1–5, Book 1, Tappi Press, Atlanta, GA, p 27

100. Anonymous (1992) No matter what you call it, chlorine free bleaching is here to stay. Pulp and Paper, Canada, 93(5): 22

101. Young J (1994) Enzone bleaching, enzynk deinking advance to pilot plant trial, Pulp and Paper, Nov. p 81

102. Lavielle P (1992) Three large scale uses of xylanases in kraft pulp bleaching, SPCI-ATICELCA 92, European Pulp and Paper Week: New Available Technologies and Current Trends/Pulp: Pulp and Paper Maintenance, Bologna, Italy, May 19–22, p 203

103. Lavielle P (1992) Three large scale uses of xylanases in kraft pulp bleaching In: Proceedings of Pan-Pacific Pulp and Paper Technology, Tokyo, Japan, September 8–10, Part A, p 59

104. Anonymous (1991) Breakthrough in Finland. chlorine free bleaching of sulphate pulp, Skogindustri, 45(10): 10

105. Worster H (1993) Canadian mills begin chlorine free bleaching as TCF pulp market grows. Pulp and Paper 67(1), 117

106. Anonymous (1992) Development of bleaching technology in Finland, Paper Puu, 74(2): 102

Biotransformation for L-Ephedrine Production

P.L. Rogers, H.S. Shin and B. Wang
Department of Biotechnology, University of New South Wales, Sydney 2052,
Australia

1 Introduction 34
2 Factors Affecting L-phenylacetylcarbinol (L-PAC) Production 36
  2.1 Pyruvate Decarboxylase (PDC) Activity 36
  2.2 Metabolic Status of Yeast 37
  2.3 Benzaldehyde Concentration 37
  2.4 End-product Inhibition 41
3 Biotransformation Using Yeasts – Free Cells 41
  3.1 Fed-Batch Process 41
  3.2 Continuous Process 42
4 Biotransformation Using Yeasts – Immobilised Cells 47
5 Biotransformation Using Purified PDC – Free Enzyme 47
  5.1 Characteristics of Purified PDC 49
  5.2 Factors Influencing Biotransformation Kinetics 49
  5.3 Kinetic Analysis 52
6 Biotransformation Using Purified PDC–Immobilised Enzyme 54
  6.1 PDC Immobilisation 55
  6.2 Factors Influencing Biotransformation Kinetics 55
  6.3 Continuous Biotransformation 55
7 Discussion and Conclusions 56
8 References 58

L-ephedrine is widely used in pharmaceutical preparations as a decongestant and anti-asthmatic compound. One of the key intermediates in its production is L-phenylacetylcarbinol (L-PAC) which can be obtained either from plants (*Ephedra* sp.), chemical synthesis involving resolution of a racemic mixture, or by biotransformation of benzaldehyde using various yeasts.

In the present review, recent significant improvements in the microbial biotransformation are assessed for both fed-batch and continuous processes using free and immobilised yeasts. From previous fed-batch culture data, maximal levels of L-PAC of $10-12 \, g \, l^{-1}$ were reported with yields of 55–60% theoretical based on benzaldehyde. However, recently concentrations of more than $22 \, g \, l^{-1}$ have been obtained using a wild-type strain of *Candida utilis*. This has been achieved through optimal control of yeast metabolism (via microprocessor control of the respiratory quotient, RQ) in order to enhance substrate pyruvate production and induce pyruvate decarboxylase (PDC) activity. Processes involving purified PDC have also been evaluated and it has been demonstrated that L-PAC levels up to $28 \, g \, l^{-1}$ can be obtained with yields of 90–95% theoretical based on the benzaldehyde added. In the review the advantages and disadvantages of the various strategies for the microbial and enzymatic production of L-PAC are compared.

In view of the increasing interest in microbial biotransformations, L-PAC production provides an interesting example of enhancement through on-line control of a process involving both toxic substrate (benzaldehyde) and end-product (L-PAC, benzyl alcohol) inhibition.

Advances in Biochemical Engineering
Biotechnology, Vol. 56
Managing Editor: Th. Scheper
© Springer-Verlag Berlin Heidelberg 1997

1 Introduction

The transformation of organic compounds using biocatalysts – enzymes, cell organelles or whole cells – are important processes in organic synthesis, and have been widely used in the production of steroids, antibiotics, vitamins and other high-value products [1]. The advantages of such biotransformations are that they are reaction-specific processes (e.g. condensation, hydroxylation, esterification, decarboxylation, etc.), and can involve high degrees of regio- and stereospecificity. In addition, in most cases relatively mild reaction conditions are used [2]. As well as these advantages some problem areas have been identified. These include low substrate and product solubilities in the aqueous phase, substrate and product toxicities, together with low yields and low productivities particularly with whole cell biotransformations. As a result there has been considerable recent research on the enhancement of biotransformation kinetics by the use of two-phase systems [3, 4, 5], biotransformation in a 'micro-aqueous' environment in the presence of an organic solvent [6, 7, 8], the use of reverse micelles [9, 10] and the application of supercritical fluid extraction for product removal [11, 12].

Microbial biotransformation for the production of biologically active chiral compounds (L-form), which are important building blocks in the pharmaceutical industry, is a field which has grown significantly in recent times. This has been largely due to the toxicological problems which have been encountered with the pharmaceutical use of racemic mixtures [13].

In the present review, various strategies are assessed for a biotransformation process which involves the conversion of benzaldehyde to L-phenylacetylcarbinol (L-PAC), an intermediate in the production of L-ephedrine and related pseudoephedrines. This can be considered as an interesting biotransformation process for evaluation, as it includes both toxic substrate and end-product inhibition.

L-ephedrine is a natural plant alkaloid isolated originally from the dried young branches of *Ephedra*, a plant with interesting pharmacological activities. Extracts of *Ephedra* sp., particularly *Ephedra sinica*, *E.equisetina* and *E.distachya* commonly called 'Ma Huang' in China, have been used for several thousand years as folk remedies for inducing sweat, soothing breath and easing excretion of urine. The active ingredient of these extracts, L-ephedrine, was first isolated in 1855, and international interest in this drug was stimulated by the classical investigations of Chen & Schmidt in 1930, who reported on its cardiovascular effects and its similarity to epinephrine [14].

Ephedrine, known chemically as 1-methylamino-ethyl-benzyl alcohol, or 2-methylamino-1-phenyl-1-propanol, contains two asymmetric carbon atoms, so that there are four optically active forms (Fig. 1), of which the (L-) isomers are used clinically. These four isomers occur naturally in *Ephedra* plants, and can be extracted with alcohol and benzene. Purified ephedrine is obtained as odourless, colourless crystals or as a white crystalline powder with bitter taste. Ephedrine

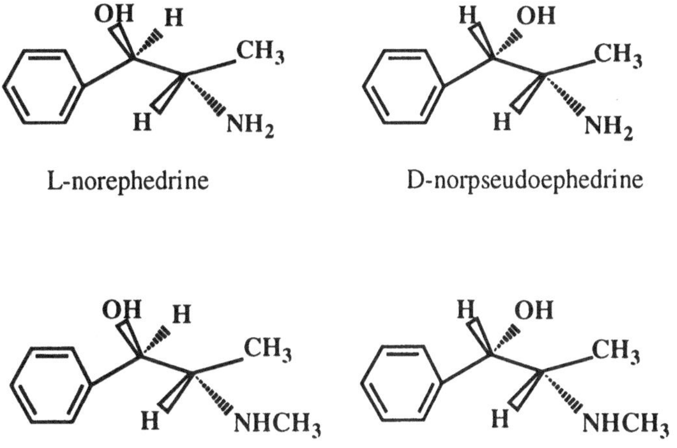

L-norephedrine D-norpseudoephedrine

L-ephedrine D-pseudoephedrine

Fig. 1. The structural formulae for ephedrine and its analogs [15]

and pseudoephedrine are very stable. A solution of ephedrine hydrochloride sealed for six years showed neither oxidation nor loss of activity [15, 16]. Its main use pharmacologically is as a decongestant or anti-asthmatic compound, although recent reports have indicated its potential in obesity control [17].

Three production methods have been used for L-ephedrine: traditional extraction from plant species of *Ephedra*, a synthetic chemical process involving resolution of the racemic mixture, and a process which involves the biotransformation of benzaldehyde to L-PAC by various species of yeasts followed by reductive amination [18, 19].

The biotransformation process which involves the condensation of an 'active acetaldehyde' (from pyruvic acid) and benzaldehyde is shown in Fig. 2. The production of the L-PAC is catalysed by the enzyme pyruvate decarboxylase (PDC), and can be associated with benzyl alcohol by-product formation, due to the activity of alcohol dehydrogenase(s) (ADH) and/or other non-specific oxidoreductases. Some traces of benzoic acid as a by-product have also been reported.

Previous studies have reported concentrations of 10–12 $g\,l^{-1}$ L-PAC in batch culture with free cells [20, 21] and with immobilised yeast [22, 23]. The addition of cyclodextrins (particularly β-cyclodextrin) was shown to enhance L-PAC production with immobilised *Saccharomyces cerevisiae* [24] although again maximal concentrations of 12 $g\,l^{-1}$ L-PAC were reported. Evaluation of a large number of yeast strains by Netrval and Vojtisek [19] identified strains of *Saccharomyces*, *Candida* and *Hansenula* sp. which were capable of significant L-PAC production in shake-flask cultures (up to a maximum of 6.3 $g\,l^{-1}$). Strain improvement studies to develop acetaldehyde and L-ephedrine resistant mutants have been reported by Seely et al. [25] with mutants capable of increased L-PAC levels compared to wild-type strains. However, maximal L-PAC levels did not

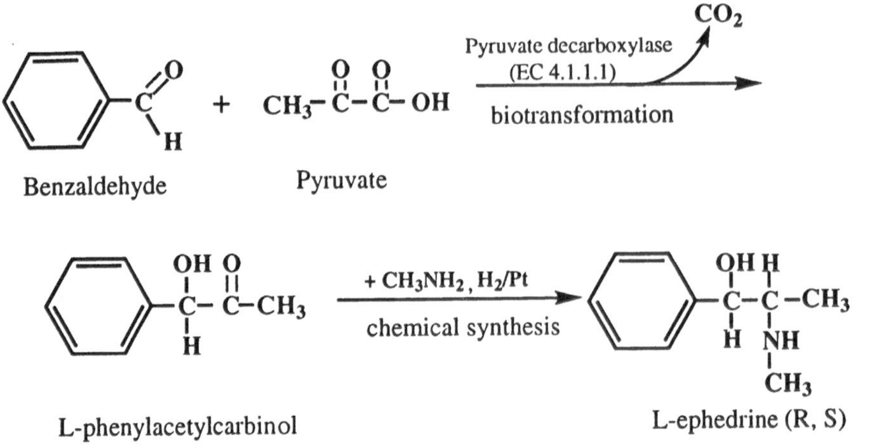

Fig. 2. Mechanism of L-PAC and ephedrine production

exceed $10 \, g \, l^{-1}$. In further fundamental investigations into the biotransformation of aromatic substrates by yeasts, other authors have indicated that oxidoreductases other than alcohol dehydrogenases may be involved in benzyl alcohol by-product formation [26–31].

Current commercial practice involves a fed-batch process with fermentative yeast growth on sugars to produce biomass, pyruvic acid and induce PDC activity. The growth phase is followed by biotransformation, with the further addition of sugars and the programmed feeding of benzaldehyde to maximise L-PAC production. Cessation of L-PAC formation occurs as a result of the following factors acting either together or independently:

(a) reduction of PDC activity due to benzaldehyde inhibition;
(b) pyruvic acid limitation at end of biotransformation phase;
(c) cell viability loss due to extended exposure to benzaldehyde and/or increasing concentrations of benzyl alcohol and L-PAC.

2 Factors Affecting L-phenylacetylcarbinol (L-PAC) Production

2.1 Pyruvate Decarboxylase (PDC) Activity

Although L-PAC production depends on PDC activity, little research has been published on the induction and deactivation of this enzyme during the biotransformation process. In our research, studies have been carried out on a strain of *Candida utilis* selected for its ability to tolerate benzaldehyde and produce relatively high levels of L-PAC. The yeast has been grown in batch culture under partial fermentation conditions to produce biomass and also induce PDC activity. As shown in Fig. 3, after 18–20 h the PDC activity reached a level of $0.9\,U\,mg\,protein^{-1}$, while the ADH activity was $1.4\,U\,mg\,protein^{-1}$. The fermentative nature of the metabolism was illustrated by ethanol accumulation of $35\,g\,l^{-1}$. As indicated earlier, the enhanced PDC activity should be conducive to rapid L-PAC production, while the higher ADH activity may result in more benzyl alcohol.

2.2 Metabolic Status of Yeast

In a study to develop strategies for on-line control of the L-PAC process, the effect of the respiratory quotient (RQ) on the production of L-PAC has been investigated by our group. An RQ value of 1.0 for growth on glucose corresponds to full respiratory growth while RQ values greater than 1.0 indicate fermentative metabolism. Increasing fermentation results in increasing RQ values. Data using cells of *C.utilis* grown in continuous culture at controlled RQ values of 1.0, 1.9 and 4.4 are shown in Fig. 4. The biotransformations were carried out in shake flasks at 30°C, pH = 6.0 at various initial concentrations of benzaldehyde ($0.5-4.0\,g\,l^{-1}$). Initial specific rates L-PAC (q_{PAC}) and benzyl alcohol production (q_{BA}) were estimated from the kinetic data. From Fig. 4 it is evident that full respiratory growth (RQ = 1.0) resulted in a low specific rate of L-PAC and a high specific rate of benzyl alcohol production. Increasing RQ had a significant effect in improving L-PAC production rates and yields (65–70% theoretical at higher RQ values). For the biotransformation process, an RQ value of 4–5 was optimal for the yeast growth phase as this corresponded to an RQ which produced sufficient yeast biomass for rapid biotransformation, while enhancing L-PAC production rates and yields. Through the use of on-line gas analysers for O_2 (paramagnetic) and CO_2 (infrared) measurements, interfaced to a microprocessor for calculation of RQ and subsequent stirrer speed adjustment, an effective feedback control strategy for RQ has been developed.

2.3 Benzaldehyde Concentration

The effect of benzaldehyde on L-PAC production by *S.cerevisiae* has been investigated by Gupta et al. [32] and Agarwal et al. [33]. The latter authors reported that once the benzaldehyde concentration increased above 16 mM ($1.7\,g\,l^{-1}$), the specific rate of L-PAC production decreased, and beyond 20 mM ($2.1\,g\,l^{-1}$) it was inhibited completely. When the residual benzaldehyde declined below 4 mM ($0.4\,g\,l^{-1}$), the formation of benzyl alcohol was predominant over L-PAC. They estimated that the optimum benzaldehyde concentration range was 4–16 mM. Further studies [34] cited the optimum concentration for L-PAC production as 10 mM ($1.1\,g\,l^{-1}$).

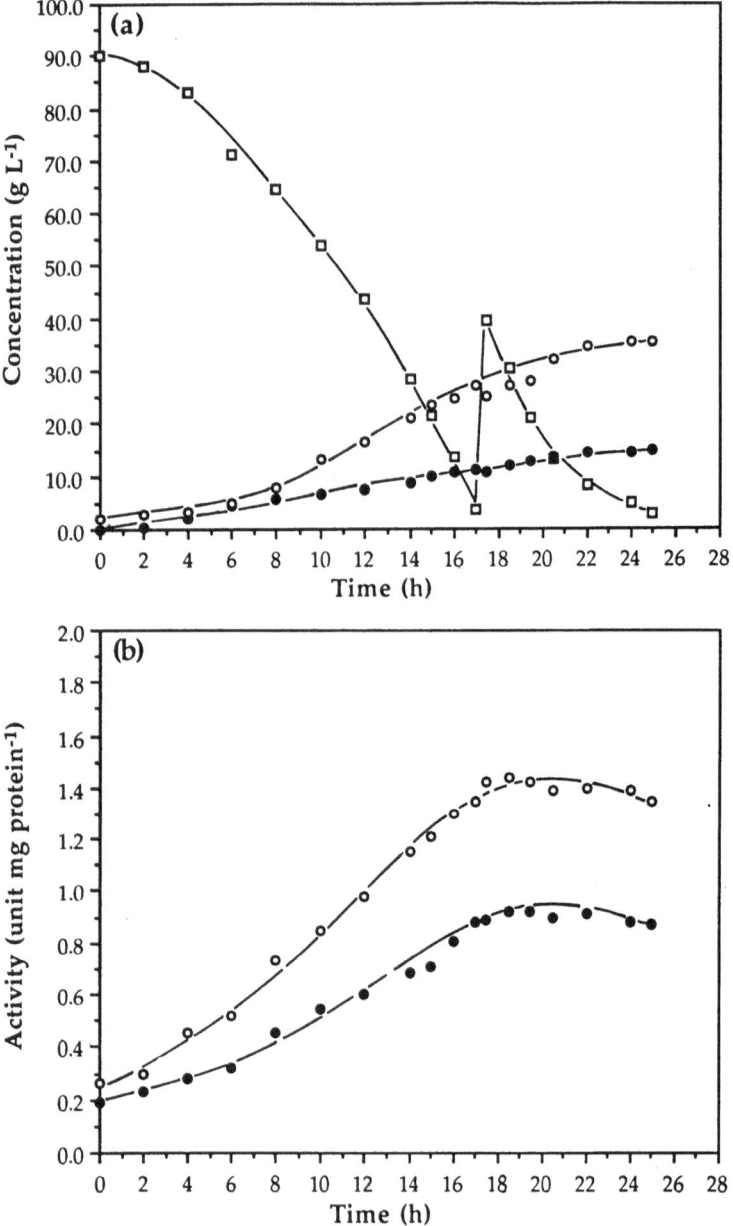

Fig. 3. Kinetics of growth and enzyme profiles of *Candida utilis*. **a** Cell growth: (●) biomass, (o) ethanol, (□) glucose. **b** Enzyme profiles: (●) PDC, (o) ADH

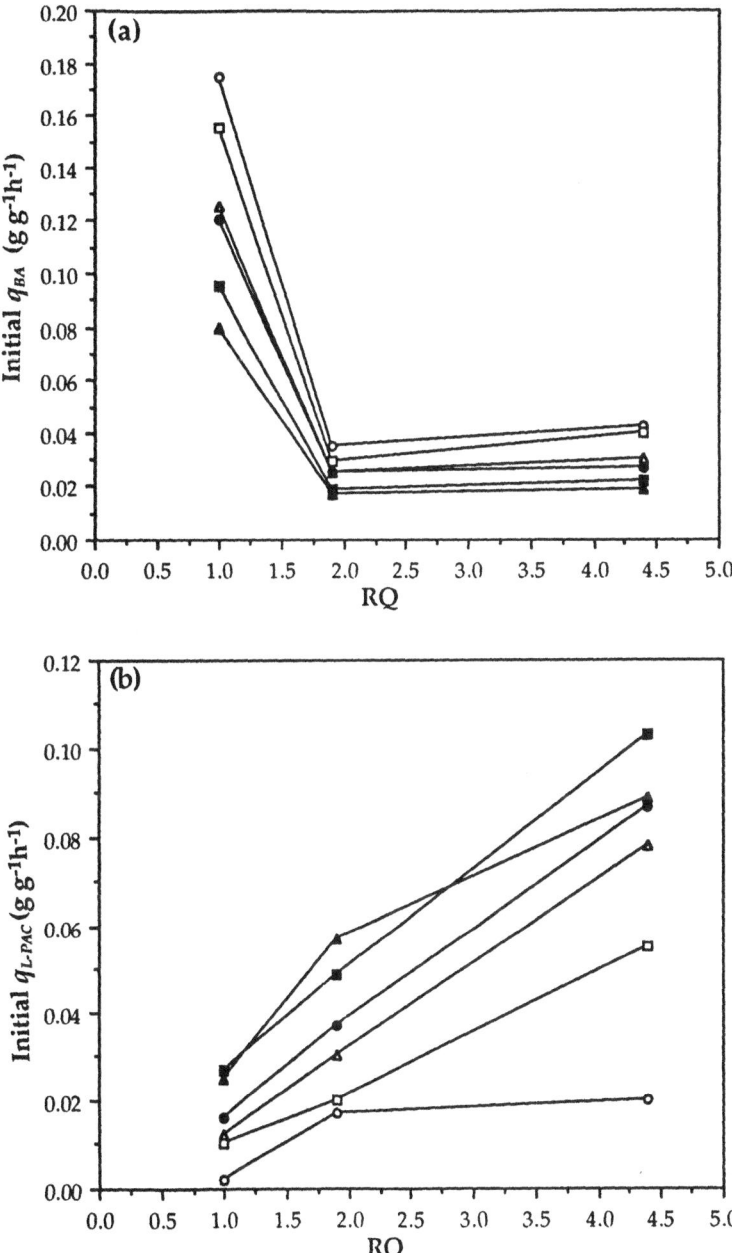

Fig. 4a–c. Effect of RQ and benzaldehyde concentrations on **a** initial specific rate of L-PAC (q_{PAC}) production, **b** specific rate of benzyl alcohol (q_{BA}) production, and **c** final L-PAC conversion efficiency. Benzaldehyde concentrations: (o) 0.5, (□) 1.0, (△) 1.5, (●) 2.0, (■) 3.0, (▲) $4.0\,\mathrm{g\,l^{-1}}$

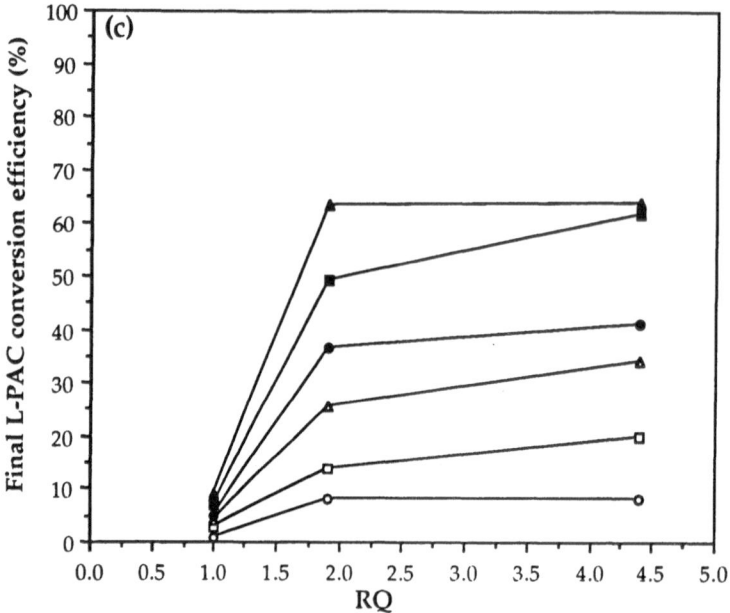

Fig. 4. (Continued)

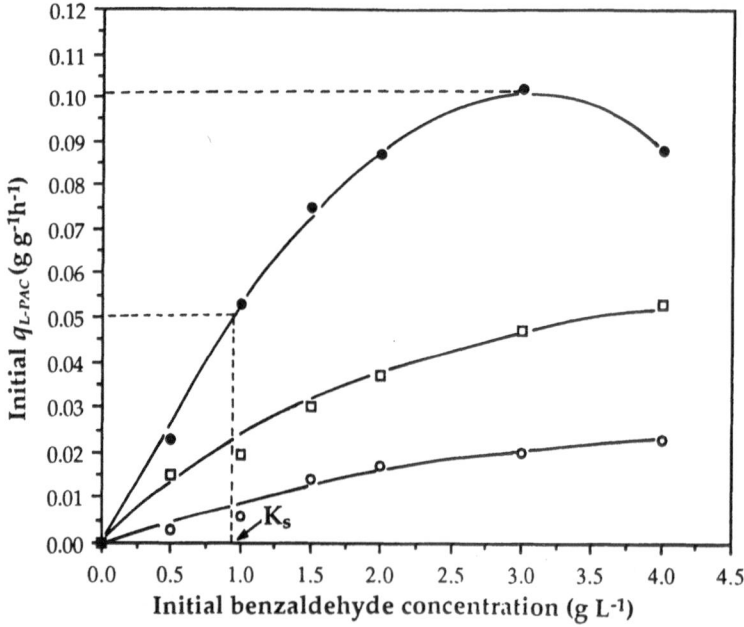

Fig. 5. Effect of benzaldehyde concentration on the specific rate of L-PAC production by *C.utilis*. Cells grown in continuous culture at T = 30 °C, pH = 5.2, D = 0.15 h^{-1} and RQ values (o) 1.0, (□) 1.9, (●) 4.4

In recent studies by our group it has been found with *C.utilis* that cessation of growth occurred for benzaldehyde concentrations above $1.0 \, g \, l^{-1}$. A growth inhibition constant for benzaldehyde was estimated at $0.30 \, g \, l^{-1}$ [unpublished data]. Using initial rate analysis, the effect of benzaldehyde on L-PAC production has been investigated. An analysis of the data for a range of initial benzaldehyde concentrations $(0.5–4.0 \, g \, l^{-1})$ is shown in Fig. 5. From the analysis, the maximum specific rate of L-PAC production (q_{PAC}) was estimated as $0.10 \, gg^{-1} h^{-1}$ for RQ = 4.4, the K_S value (saturation constant) was $1.0 \, g \, l^{-1}$ benzaldehyde and concentrations of benzaldehyde in excess of $3.0 \, g \, l^{-1}$ caused a significant decrease in q_{PAC}. In studies of Tripathi et al. [34] with a strain of *S.cerevisiae*, a maximum q_{PAC} value of $0.07 \, gg^{-1} h^{-1}$ was reported.

2.4 End Product Inhibition

The main products which are likely to influence yeast growth and metabolism as well as L-PAC production are: ethanol produced by fermentative metabolism, benzyl alcohol by-product and the L-PAC product itself. It is possible also that acetaldehyde or benzoic acid accumulation might influence L-PAC production, although only very low levels were detected during biotransformation with free cells of *C.utilis*

In this context, Seely et al. [25] have mutagenized strains of *S.cerevisiae* and *C.flareri* using classical methods, eg. *N*-methyl-*N'*-nitro-*N*-nitrosoguanidine (NTG), ultra-violet light (UV) and gamma rays and have selected mutants with increased resistance to acetaldehyde and L-ephedrine. Strains were isolated capable of producing L-PAC to higher levels than parent cultures, and final concentrations to a maximum of L-PAC of $9.9 \, g \, l^{-1}$ were reported.

Studies in our laboratory on a strain of *C.utilis* have indicated the potential for significant end-product and by-product inhibition of growth. Product inhibition constants for growth (K_P) have been estimated as $4.1 \, g \, l^{-1}$ L-PAC, $5.0 \, g \, l^{-1}$ benzyl alcohol and $39 \, g \, l^{-1}$ ethanol [unpublished data]. The effect of these products on the specific rate of L-PAC production (q_{PAC}) for the free cells are yet to be determined.

3 Biotransformation Using Yeasts – Free Cells

3.1 Fed-Batch Process

A fed-batch process for L-PAC production can be subdivided into three basic phases:

(i) growth of the yeast under partial fermentation conditions to produce sufficient biomass for the biotransformation process, and also to facilitate the accumulation of pyruvate for subsequent L-PAC production;

(ii) optimal induction of PDC for biotransformation – however PDC induction in an increasingly fermentative environment results in conversion of pyruvate to ethanol via acetaldehyde, and an increase in alcohol dehydrogenases which may enhance benzyl alcohol formation;

(iii) programmed feeding of benzaldehyde to maintain concentrations of $1-2 \, g \, l^{-1}$ and the subsequent production of L-PAC.

As discussed earlier, L-PAC concentrations of $10-12 \, g \, l^{-1}$ have been reported in the literature using various species of yeasts. However, it is only recently that higher levels have been found. Presumably pyruvate depletion occurred in many of the earlier studies and insufficient attention was paid to initial pyruvate accumulation, PDC activity and the metabolic status of the yeast. Some studies [20, 28] evaluated the addition of pyruvate and acetaldehyde to achieve higher L-PAC levels; however the effects were variable and in many cases no enhancement was found.

In a recent study by our group, L-PAC levels up to $22 \, g \, l^{-1}$ have been achieved through the optimal control of metabolism of a strain of *C.utilis* (viz. RQ = 4–5) and the sustained feeding of benzaldehyde while maintaining a concentration of $1-2 \, g \, l^{-1}$ in the bioreactor [35]. The results of a typical experiment in a controlled bioreactor (T = 20 °C, pH = 6.0) using a fully-defined medium are shown in Fig. 6. Prior to PDC activation, pyruvic acid levels were typically $10-15 \, g \, l^{-1}$ with PDC activity low (0.2 U mg protein^{-1}). After cell growth and pyruvate accumulation (total 22 h) with glucose addition to maintain active metabolism, PDC activity was enhanced (to greater than 1.0 U mg protein^{-1}). Benzaldehyde feeding was initiated and resulted in subsequent L-PAC production. As shown in Fig. 6, accumulation of benzyl alcohol occurred to $4 \, g \, l^{-1}$ and L-PAC reached $22 \, g \, l^{-1}$. Biotransformation finally ceased due to pyruvate depletion accompanied by declining PDC activity. Cell viability studies after 14 h biotransformation indicated 100% viability loss. The productivity for the biotransformation phase was $1.6 \, g \, l^{-1} h^{-1}$ with a yield of 65% theoretical based on the benzaldehyde utilised. No benzoic acid was detected.

3.2 Continuous Process

Continuous processes are often cited as having the advantages of higher productivities, easier process control and compatible operation with subsequent downstream product recovery. However, there are relatively few reports on the application of free cells in continuous culture to biotransformation processes, although Rohner et al. [36, 37] have reported on the successful application of chemostat cultures of *S.cerevisiae* for the stereospecific reduction of acetoacetic acid esters to the 3-(5)-hydroxy-butanoic acid esters. A continuous single-stage steady-state production system was found to be superior to the pulse-batch and fed-batch systems in terms of product optical purity, biomass concentration and production rates. It was established also in this process that the reduced product

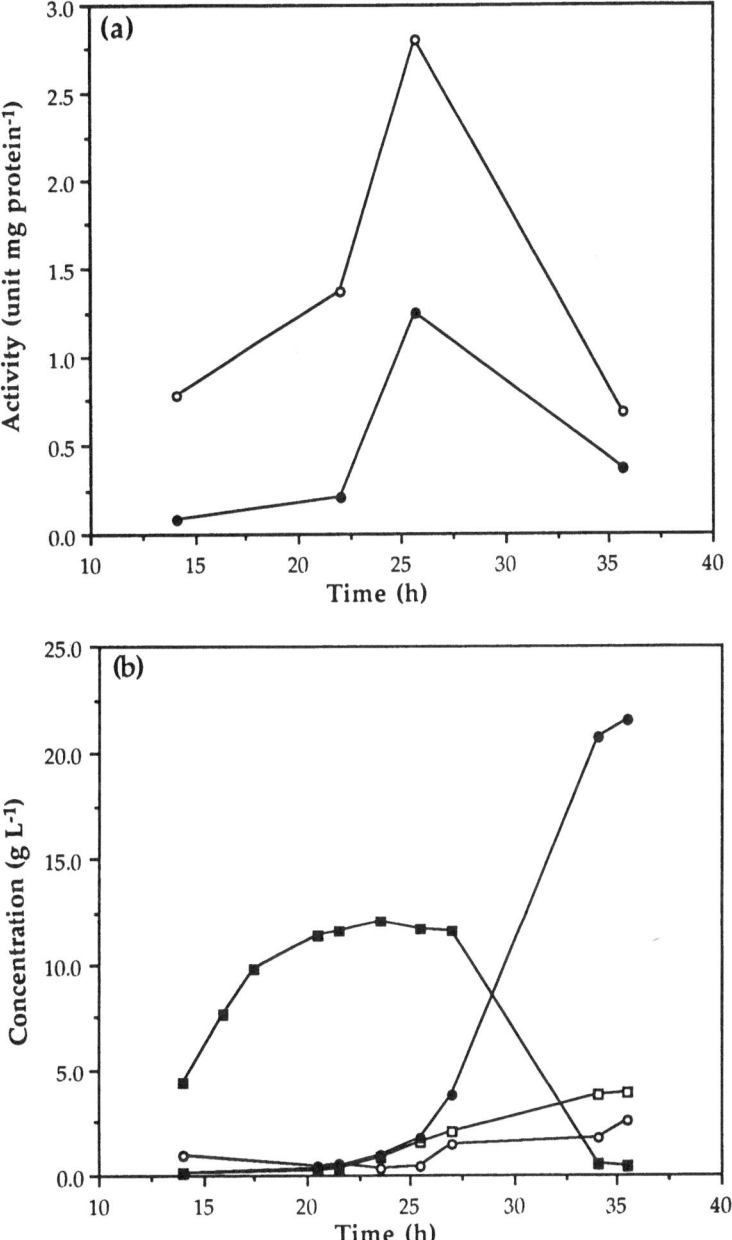

Fig. 6a, b. Fed-batch biotransformation of benzaldehyde to L-PAC by *C.utilis*. **a** Enzyme profiles: (●) PDC, (○) ADH, **b** Biotransformation kinetics: (○) benzaldehyde, (□) benzyl alcohol, (■) pyruvate, (●) L-PAC

44 P.L. Rogers, H.S. Shin and B. Wang

(3-(5)-hydroxybutanoic acid ester) did not inhibit the cells [36] although the starting material (acetoacetic acid ester) influenced both the respiration and cell physiology of *S.cerevisiae*.

With L-PAC production, both the substrate (benzaldehyde) and the product (L-PAC) have significant inhibitory effects on yeast metabolism and PDC activity. These factors necessitate the design of a more complex continuous culture process, and suggest one involving multistage operation and later stage benzaldehyde feeding in order to sustain steady-state conditions. Other workers have reported on the use of a single-stage continuous culture for the production of active yeast cells for L-PAC production [34], however the biotransformation of benzaldehyde was carried out in batch culture.

In recent studies in our laboratories a detailed evaluation of various continuous culture systems has been carried out, and the results of this research can be summarised as follows:

(i) A single-stage continuous culture operating as a glucose-limited chemostat with continuous feeding of benzaldehyde was not suitable for the biotransformation process due to the strongly inhibitory effects of benzaldehyde on cell growth.

(ii) For a two-stage system, partially fermentative conditions can be established in the first stage to provide for pyruvic acid accumulation and PDC induction, while benzaldehyde can be added to the second stage to facilitate the biotransformation. However, it was found that the biomass yield under the partially fermentative conditions in the first stage was too low to produce sufficient yeast cells for rapid L-PAC production in the second stage.

(iii) To overcome the disadvantages of the single- and two-stage processes, a three-stage system was designed as follows:

Stage 1. A fully aerobic stage (RQ = 1) designed to give yields to *C.utilis* of $Y_{x/s} = 0.45$–0.50 gg^{-1} glucose, close to the theoretical maximum biomass yield.

Stage 2. A partially fermentative stage (RQ = 4–5) designed to increase PDC activity as well as providing for some accumulation of pyruvic acid. A supplementary feed of glucose was supplied to this second stage.

Stage 3. A continuous biotransformation stage was established with benzaldehyde addition at various feed rates. Low level glucose feeding was also supplied to this stage to provide substrate for continuing yeast metabolic activity.

A diagram of the three-stage process together with a typical set of operating conditions is shown in Fig. 7.

This system was evaluated with a range of benzaldehyde and supplementary glucose feed rates using a defined medium containing 60 g l^{-1} glucose to the first stage. Under fully aerobic conditions (DO > 20% air saturation), it was established that close to 30 g l^{-1} biomass was produced in the first stage (RQ = 1). PDC induction up to 0.50 U mg protein^{-1} and pyruvate accumulation to 2–3

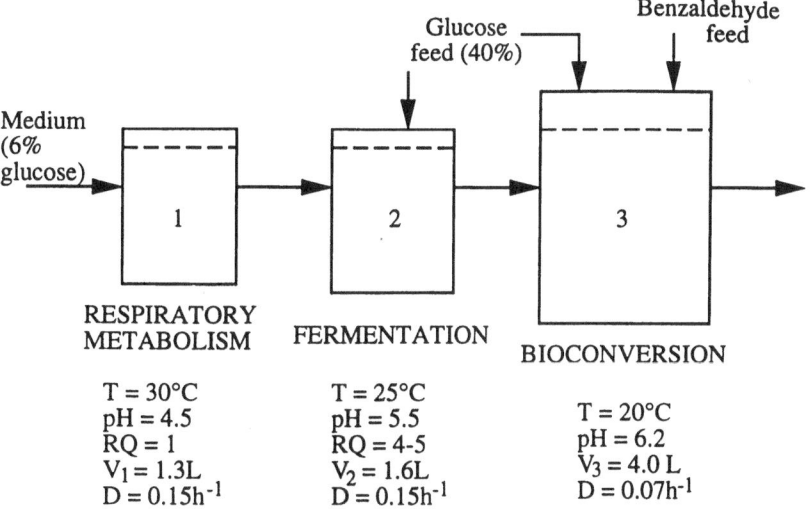

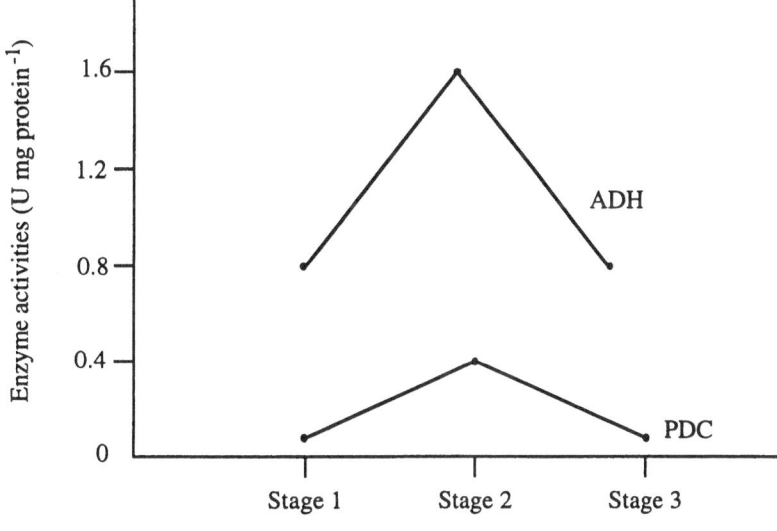

Operating characteristics for 3-stage continuous
process for PAC production

Fig. 7. Operating conditions for a three-stage continuous biotransformationprocess, showing typical enzyme activities for each stage

gl^{-1} occurred during Stage 2, as well as the formation of 45–50 gl^{-1} ethanol. The ethanol has been shown to be beneficial for enhancing the solubility of benzaldehyde during Stage 3; however it represented a significant consumption of glucose. As shown in Fig. 8, when the benzaldehyde feeding rate was increased

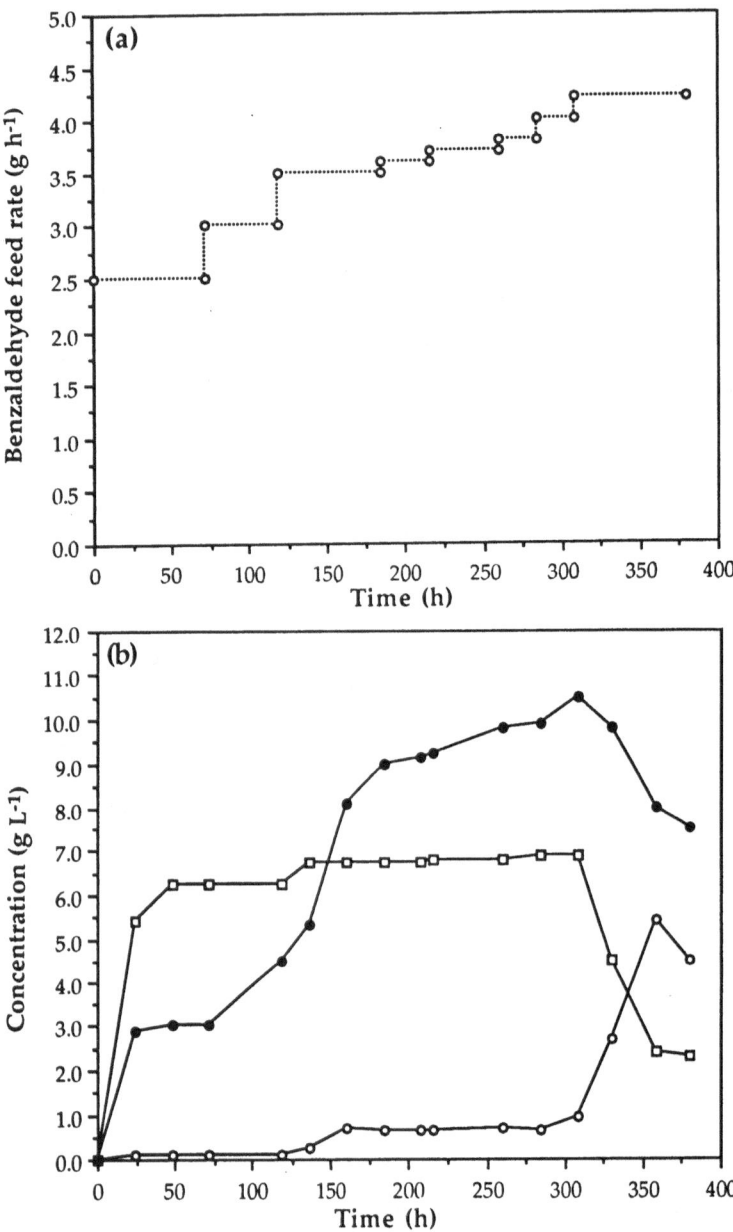

Fig. 8a, b. Effect of increased benzaldehyde feeding on L-PAC production in a three-stage process, **a** Benzaldehyde feed rate; **b** concentrations: (o) benzaldehyde, (□) benzyl alcohol, (●) L-PAC

incrementally during 300 h, the L-PAC concentration increased to 10.6 g l^{-1}. Benzaldehyde feeding rates of greater than 4 g h^{-1} to a controlled fermentor of 4 l working volume during Stage 3 resulted in a decline in L-PAC levels and the accumulation of benzaldehyde. Under these conditions, the PDC activity declined to less than 0.05 U mg protein^{-1}. At the maximal steady-state L-PAC concentration of 10.6 g l^{-1}, the productivity, based on the total volume of the three fermentors, was 0.44 g l^{-1} h^{-1} with a yield of 0.80 gg^{-1} (56% theoretical) based on the benzaldehyde used.

4 Biotransformations Using Yeasts – Immobilised Cells

It has been reported that cell immobilisation of yeasts is able to reduce the toxic effects of benzaldehyde by virtue of diffusional limitations and the toxic substrate gradients that are established within the immobilising matrix [22, 23, 38].

Recent studies by our group using cells of *C.utilis* immobilised within calcium alginate beads of 2–3 mm diameter have confirmed this enhanced resistance to benzaldehyde [39], with the immobilised cells producing higher L-PAC levels than free cells in shake flask experiments. In experiments with the programmed feeding of benzaldehyde in a controlled bioreactor (T = 20 °C, pH = 5.0), under similar conditions to those described previously with free cells, the final L-PAC concentration was 15 g l^{-1} [39]. The PDC activity of the cells in the beads which was initially at 0.65 U mg protein^{-1} declined to 0.2 U mg protein^{-1}, with cessation of L-PAC production resulting from pyruvate depletion. Electron microscope pictures taken at the end of the biotransformation (Fig. 9) indicated that major cell wall damage had occurred within the calcium alginate beads. The reason that the final L-PAC level of 15 g l^{-1} with immobilised cells was significantly less than the 22 g l^{-1} achieved with the free-cell fed-batch system is that better control of yeast metabolism (via RQ) in the latter case provided for greater pyruvate accumulation. Levels of 10–15 g l^{-1} were achieved with the free cells, while for the immobilised cell system the maximum concentration was only 5 g l^{-1}.

When this evaluation with immobilised *C.utilis* was extended to a continuous bioreactor, the steady-state L-PAC levels were low (no more than 4 g l^{-1} in sustained operation). These results were similar to those of Mahmoud et al. [22, 38] who reported on a semicontinuous immobilised cell process. These authors found with immobilised *S.cerevisiae* that L-PAC production in the first and second cycles (with an intervening reactivation period of 24 h) could reach 4.5 g l^{-1}. Attempts to extend the process to three or more cycles resulted in rupturing of the cells/beads in the continuing presence of benzaldehyde.

5 Biotransformation with Purified PDC – Free Enzyme

One of the problems of using yeast whole cells for the biotransformation process is that considerable amounts of benzaldehyde are converted to the unwanted

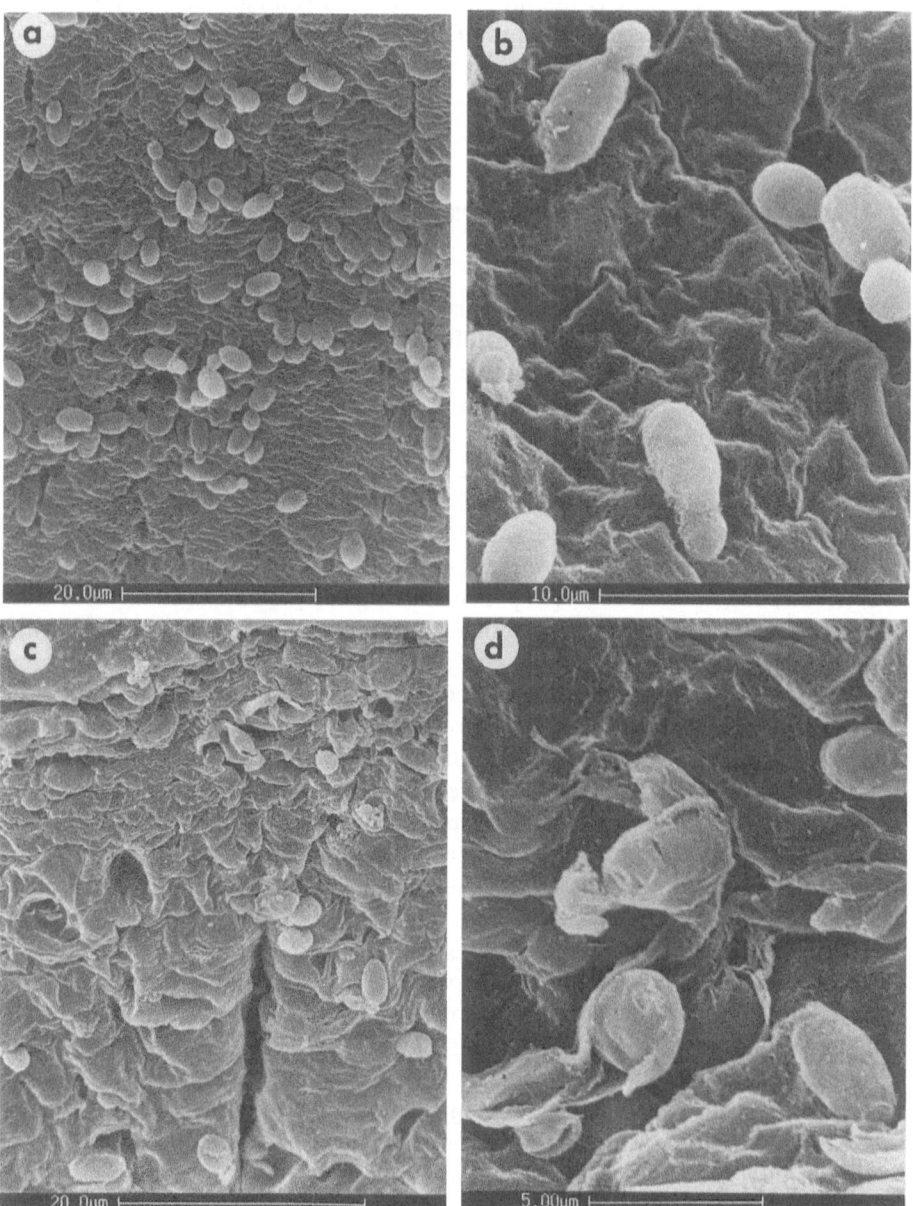

Fig. 9a–d. Comparison of electron microscope pictures of immobilised cells of *C.Utilis* in calcium alginate beads. **a, b** Before exposure to benzaldehyde, **c, d** following prolonged exposure to 1–2 g l^{-1} benzaldehyde. The scales are given on the four photographs

by-product, benzyl alcohol. The use of purified PDC offers the possibility of overcoming this problem. However, in the latter case, pyruvate will have to be supplied as a substrate and there is the further likelihood that the pyruvate added will be removed via decarboxylation to acetaldehyde and condensation to acetoin.

Bringer-Meyer and Sahm [40] have investigated the use of purified PDC from *Zymomonas mobilis* and *S.carlsbergensis* for L-PAC production, and demonstrated that the PDC from *Z.mobilis* is unsuitable for the biotransformation due to its low affinity for benzaldehyde and because of significant substrate inhibition effects. However, the substrate levels and consequent L-PAC concentrations achieved with *S.carlsbergensis* were relatively low in this investigation.

To provide an effective comparison between the yeast biotransformation process and one which involves the purified enzyme, it is necessary to evaluate similar operating conditions and substrate concentrations. Details of our recent study [41] are as follows.

5.1 Characteristics of Purified PDC

Purified PDC was prepared from cells of *C.utilis* growing under partially fermentative conditions in a 100-l fermentor at a constant temperature of 25 °C and a pH of 6.0. When the PDC activity had reached 0.9 U mg protein^{-1}, the cells were harvested, disrupted in a high pressure homogeniser and the PDC was purified by means of $(NH_4)_2SO_4$ precipitation and gel chromatography. Previous studies with PDC have demonstrated that the enzyme has a dimeric tetramer structure $(\alpha_2\beta_2)$ which dissociates in vitro into dimer and monomer subunits with the concomitant release of the cofactors thiamine pyrophosphate (TPP) and magnesium ions [42]. This dissociation, predominantly a function of pH, is affected also by the buffer species. The dissociation was found to be greater in Tris-Cl compared with phosphate and citrate buffers. As a result, in the reported study [41], the reaction mixtures contained various PDC activities and pyruvate : benzaldehyde ratios with 30 mM TPP, 0.5 mM $MgSO_4 \cdot 7H_2O$ in 40 mM phosphate buffer (pH 6.0) in order to achieve maximal enzyme stability.

As shown in Fig. 10, the K_m value of PDC for pyruvate was determined to be 2.2 mM at 25 °C and pH 6.0, with saturation at concentrations in excess of 10 mM pyruvate. The K_m value is in good agreement with other reported values for PDC from yeasts [43–45].

5.2 Factors Influencing Biotransformation Kinetics

The effects of various factors on L-PAC formation with purified PDC have been reported in detail [41] and can be summarised as follows:

(a) Higher yields of L-PAC were achieved at lower temperatures due to reduced acetaldehyde production. For this reason, a temperature of 4 °C was selected for the enzyme biotransformations.

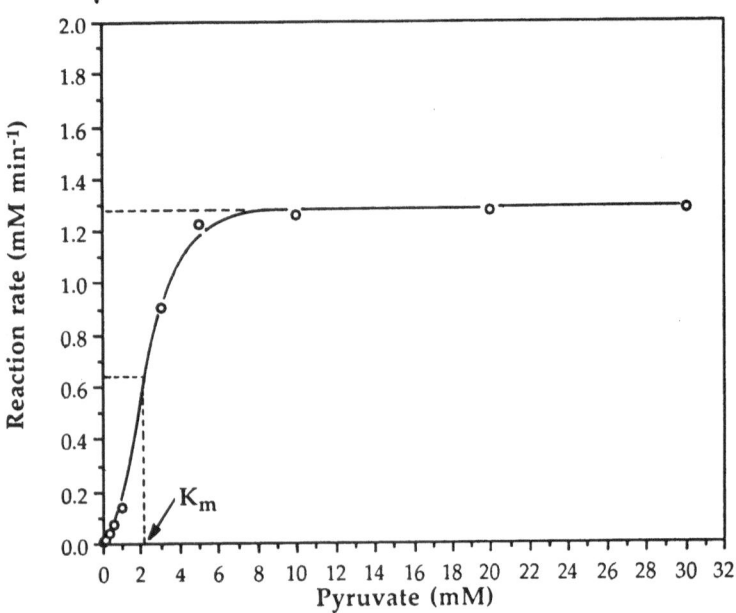

Fig. 10. Estimation of the K_m value for pyruvate for PDC from *C.utilis* at T = 25°C, pH = 6.0

(b) The optimum pH for L-PAC formation was 7.0 while that for acetaldehyde production was 6.0.
(c) As acetaldehyde is an inhibiting by-product, its role in L-PAC formation was investigated. Initial rate studies established that the inhibition constant (K_p) for acetaldehyde was of the order of 20 mM (0.9 g l^{-1}) indicating that it could play a significant role in reducing the L-PAC formation rate.
(d) The influence of an organic solvent such as ethanol in enhancing the benzaldehyde solubility and increasing the L-PAC production rate was evaluated. The selection of ethanol as a water miscible organic solvent was based on the following information: PDC has significant resistance to denaturation by ethanol up to a concentration of 3M [46], PDC has been reported to have highly hydrophobic substrate binding sites and the presence of ethanol may assist enzyme/substrate interactions [47]. Furthermore, benzaldehyde is reported to have infinite solubility in ethanol [48] compared to its very limited solubility in water (0.3 g 100 ml^{-1}). Initial rate studies showed that an increase of 30–40% in the rate could be achieved by addition of 2.0–3.0 M ethanol.

Based on the above optimal conditions, the substrate saturation constant (K_m) for benzaldehyde was determined (Fig. 11), and it is interesting to note that substrate inhibition (benzaldehyde toxicity) was evident only above concentrations of 180 mM (19.1 g l^{-1}) for the free enzyme. This compares with data for the

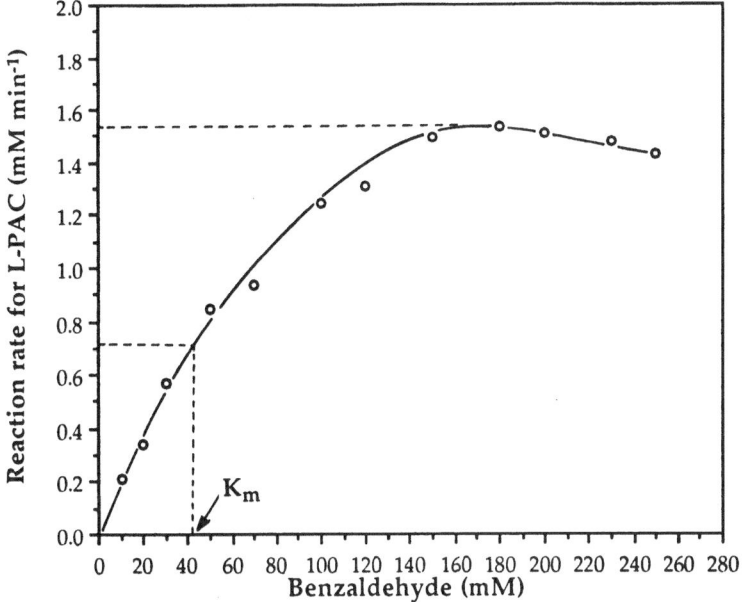

Fig. 11. Estimation of K_m value for benzaldehyde for PDC from *C.utilis* at T = 4 °C, pH = 7.0

influence of benzaldehyde on the growth of *C.utilis* where concentrations above 10 mM were found to inhibit growth completely. The K_m value for benzaldehyde substrate limitation for the free enzyme was determined to be 42 mM which compares well with values for K_m of 50 mM for PDC from *S.carlsbergensis* in the literature [40].

As well as pyruvate and benzaldehyde concentrations, it has been established that initial L-PAC formation rates are influenced significantly by PDC activity. As evident from Fig. 12, higher PDC activities resulted in higher initial rates over a wide range of benzaldehyde concentrations (up to 200 mM). Benzaldehyde toxicity was greater for the lower PDC activities. However, although the initial rates were strongly affected by PDC activity, the final L-PAC concentrations were not influenced to the same extent. There was a trend also towards increased acetaldehyde production at the higher PDC activity levels (Table 1).

The influence of the molar ratio of pyruvate: benzaldehyde is shown in Table 2. From the data, while the highest molar biotransformation yield of 97.8% (based on initial benzaldehyde) was achieved using 150 mM benzaldehyde and a 1.5 molar ratio, the highest concentration of 191 mM (28.6 g l⁻¹) L-PAC was obtained using 200 mM benzaldehyde and a 2.0 molar ratio after an 8 h biotransformation. Molar conversion yields close to 90% (based on initial pyruvate) were achieved only in the range of 150–200 mM benzaldehyde when molar ratios of 1.0 or 0.5 were used. In other situations, a higher proportion of pyruvate was converted to by-products (free acetaldehyde and acetoin) or

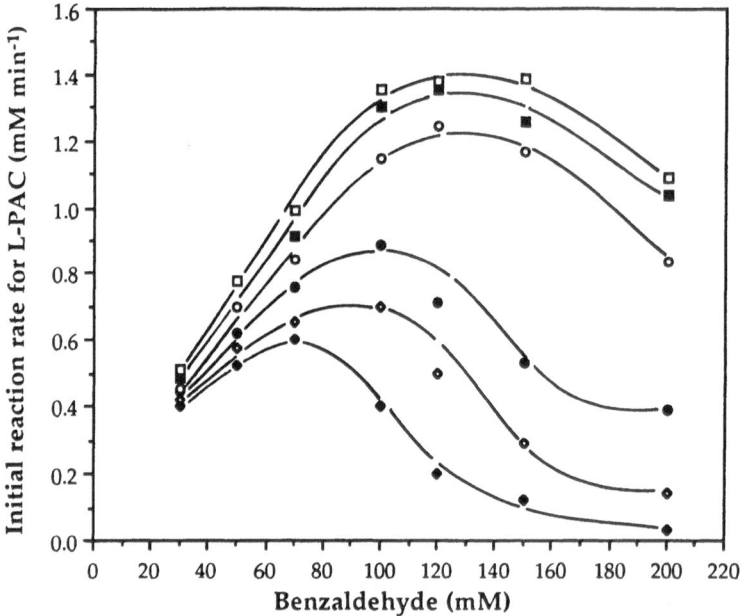

Fig. 12. Effect of PDC activity on initial rate of L-PAC formation in the presence of various concentrations of benzaldehyde and equimolar concentrations of sodium pyruvate in 40 mM phosphate buffer (pH = 7.0) containing 30 μM TPP, 0.5 mM $MgSO_4 \cdot 7H_2O$ and 2.0 M ethanol. The initial rates were determined within 30 min at 4 °C. The symbols refer to PDC activities of (■) 4.0, (□) 5.0, (●) 6.0, (o) 7.0, (■) 8.2 and (□) 10.0 unit.ml^{-1}

remained in excess concentrations in the reaction mixture. It is interesting to note also that increases in initial benzaldehyde (200–300 mM) and PDC activity (7.0–20 U ml^{-1}) did not further raise the final L-PAC concentrations, presumably because of substrate toxicity effects.

5.3 Kinetic Analysis

Detailed kinetics of a typical time course for the biotransformation of 150 mM benzaldehyde with purified PDC are shown in Fig. 13; they demonstrate the time dependence of simultaneous L-PAC, acetaldehyde and acetoin formation together with biotransformation of pyruvate and benzaldehyde. In the first 2–3 h, L-PAC formation increased rapidly, while further L-PAC formation occurred more slowly and was influenced by substrate depletion and probable product inhibition. After a 6 h incubation period, giving a maximum L-PAC concentration of 147 mM (22 g l^{-1}), a mass balance based on pyruvate indicated that of the original 225 mM pyruvate, 147 mM had contributed to L-PAC formation, 25 mM pyruvate had been converted to acetoin and 22 mM had been converted to free acetaldehyde. The residual pyruvate was 30 mM, indicating

Table 1. Final L-PAC and acetaldehyde concentrations (mM) with various PDC activities in the presence of equimolar benzaldehyde and sodium pyruvate at 4 °C and pH 7.0

PDC activity (Unit ml^{-1})	L-PAC (mM)			
	Benzaldehyde (mM)			
	30	50	100	150
4.0	19.0	34.8	48.2[a]	3.4[a]
5.0	20.1	34.8	64.5	30.0[a]
6.0	17.9	34.2	76.8	94.0
7.0	17.3	33.5	78.3	135.6
8.2	16.2	32.1	71.3	129.3
10.0	15.1	30.5	62.9	125.7

PDC activity (Unit ml^{-1})	Acetaldehyde (mM)			
	Benzaldehyde (mM)			
	30	50	100	150
4.0	6.1	9.5	2.3[a]	ND[a]
5.0	6.5	10.2	7.5	ND[a]
6.0	7.1	10.8	10.5	8.3
7.0	7.5	11.0	11.4	8.4
8.2	8.2	12.0	16.0	9.2
10.0	9.1	13.5	17.8	10.1

[a]Reaction did not proceed fully due to toxicity of benzaldehyde at low PDC activity.

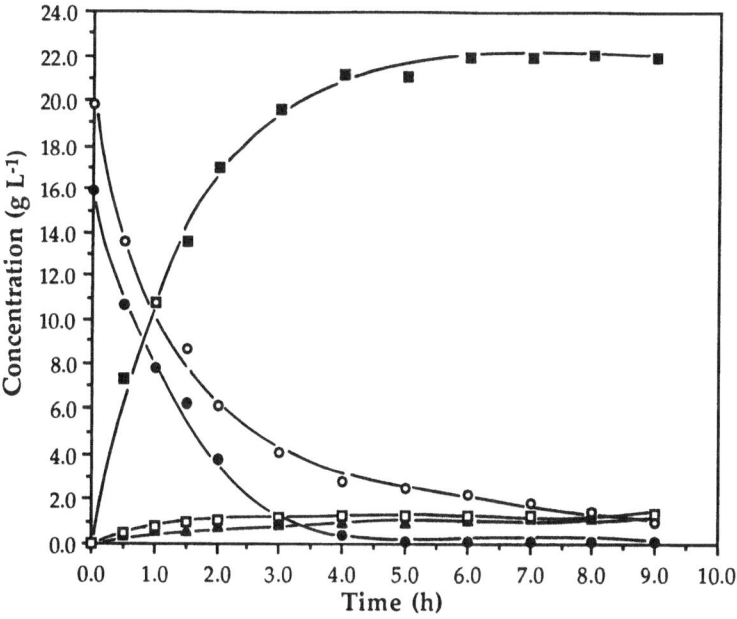

Fig. 13. Typical biotransformation kinetics with purified PDC from *C. utilis*. T = 4 °C, pH = 7.0. Concentrations (●) benzaldehyde, (○) pyruvic acid, (▲) acetaldehyde, (□) acetoin, (■) L-PAC

Table 2. L-PAC formation with various molar ratios of pyruvate to benzaldehyde at 4 °C. The reaction mixture consisted of 40 mM phosphate buffer (pH 7.0) containing 7 U ml^{-1} PDC, 2.0 M ethanol and various concentrations of sodium pyruvate and benzaldehyde

Final concentration of L-PAC (g l^{-1}) and molar conversion yields (%)

Molar ratio of pyruvate to benzaldehyde

BZ (mM)	0.5	1.0	1.2	1.5	2.0
100	5.8	11.7	14.2	14.5	14.5[a]
	77.0	78.2	78.3	64.5	48.4[b]
	38.3	78.2	94.6	96.7	96.7[c]
120	7.0	14.6	16.8	17.6	17.6
	77.2	81.1	7.6	65.2	48.9
	38.6	81.8	93.2	97.5	97.5
150	10.9	20.5	21.5	22.0	22.0
	97.0	90.6	79.6	64.9	48.9
	48.6	90.6	95.3	97.8	97.8
180	12.1	24.2	25.1	25.5	25.5
	89.5	90.0	77.5	63.0	47.2
	44.4	90.0	93.0	94.4	94.4
200	13.3	22.7	24.9	27.9	28.6
	88.5	75.6	69.1	62.0	47.6
	44.2	75.6	82.9	93.0	95.3

[a] L-PAC (g l^{-1}); expressed as g l^{-1} for comparison with literature values.
[b] $Y_{p/s}$ (mole L-PAC.mole added pyruvate^{-1}) × 100 (%).
[c] $Y_{p/s}$ (mole L-PAC.mole added benzaldehyde^{-1}) × 100 (%).

a closing mass balance based on pyruvate. With the complete utilisation of 150 mM benzaldehyde, the mass balance based on benzaldehyde conversion to L-PAC closed to within 2%. The small mass balance discrepancy was probably due to evaporative losses of benzaldehyde and/or minor experimental errors. In addition, after 6 h incubation, 20–30% of initial PDC activity still remained, indicating the potential for further biotransformation if more benzaldehyde were available.

6 Biotransformation using Purified PDC – Immobilised Enzyme

Previous studies have reported that cell and enzyme immobilisation in suitable gels and matrices could minimise substrate inhibition effects by means of diffusional limitation and the substrate gradients which exist within the immobilising material [22, 23, 38]. Furthermore, the development of an immobilised system provides the technology for long-term continuous operation, provided that enzyme stability can be maintained.

6.1 PDC Immobilisation

In our recent work, various immobilising methods have been evaluated for PDC, including adsorption on cationic exchange resins and entrapment in gel matrices [49]. It was found that entrapment in calcium polyacrylamide gel provided a higher activity than adsorption on either Amberlite IR-200 or CM-Sephadex (activities expressed as U ml^{-1} immobilising material), and that the addition of 0.2–0.3% glutaraldehyde to the polyacrylamide gel enhanced the PDC binding capacity and increased PDC activity by 40%. The 'apparent' K_m for PDC immobilised with respect to pyruvate was determined at 25 °C from initial-rate data, and a value of 3.2 mM estimated from Lineweaver-Burk analysis. This compares with 2.2 mM for the free enzyme – the higher K_m for the immobilised enzyme being consistent with mass transfer limitations within the gel.

6.2 Factors Influencing Biotransformation Kinetics

In a similar evaluation to that with free PDC, the optimal conditions for the immobilised PDC were determined. Although higher temperatures gave rise to an initial increase in the L-PAC production rate, the relatively high acetaldehyde and acetoin concentrations at 25 °C resulted in reduced final L-PAC formation. As for the free enzyme, a temperature of 4 °C was selected to maximise the L-PAC yield. The optimum pH for L-PAC with immobilised PDC was shifted slightly to more acidic conditions (pH = 6.5) compared to the free enzyme (pH = 7.0) . Ethanol was again found to enhance initial rates (due to improved benzaldehyde solubility) and a concentration of 3–4 M gave a 40% increase in initial rates compared to the control.

For a batch biotransformation process, the immobilised PDC process did not provide any real advantages compared to the free PDC system. The results were much as expected with the immobilised enzyme able to function better at higher benzaldehyde concentrations (up to 300 mM). The maximum final L-PAC concentrations were similar, viz. 27.1 g l^{-1} with 300 mM benzaldehyde and 1.5 molar ratio of pyruvate:benzaldehyde.

6.3 Continuous Biotransformation

The operating parameters for a continuous process are shown in Fig. 14 with increasing molar ratios and residence times (viz. decreasing dilution rates) resulting in the increased formation of L-PAC and by-products, acetaldehyde and acetoin. While the highest concentration of 5.3 g l^{-1} L-PAC was achieved with a 2.0 molar ratio at $D = 0.05$ h^{-1}, higher L-PAC productivities could be achieved at higher dilution rates. In an evaluation of the long-term operation of the process it was found that a gradual decline in PDC activity occurred, with the enzyme half-life estimated to be 32 days.

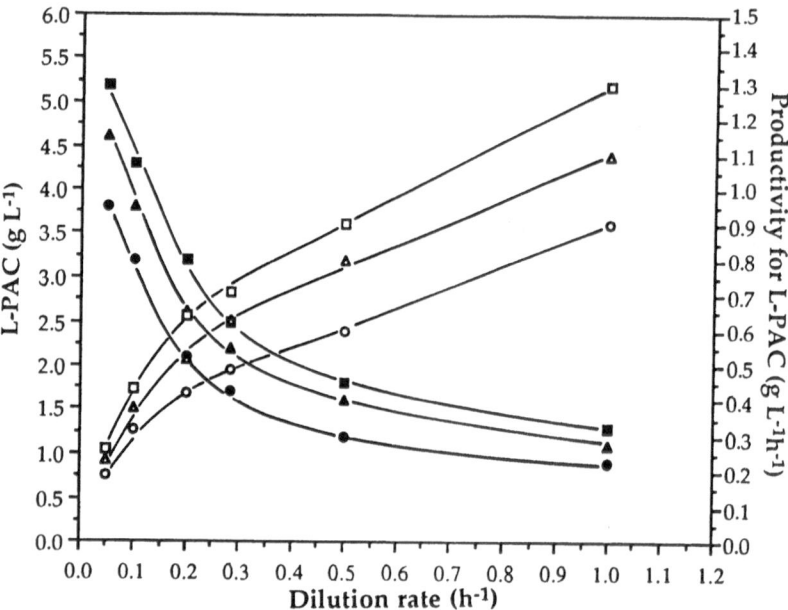

Fig. 14. Effect of dilution rate and molar ratios of pyruvate: benzaldehyde on L-PAC production using immobilised PDC in a packed-bed bioreactor. $T = 4°C$, $pH = 7.0$. L-PAC concentrations for molar ratios: (●) 1.0, (▲) 1.5, (■) 2.0; productivities for molar ratios: (o) 1.0, (△) 1.5, (□) 2.0

7 Discussion and Conclusions

A comparison of the various biotransformation processes for the conversion of benzaldehyde to L-PAC is given in Table 3. From the data it is evident that for both free and immobilised yeast, the biotransformation is a relatively low efficiency process in which there is significant diversion of benzaldehyde to benzyl alcohol and a loss of up to 30–40% due to the formation of by-products. This diversion has been attributed to the activity of alcohol dehydrogenases in yeast. However, studies by Nikolova and Ward [29] using mutant strains of *S.cerevisiae* that lack ADH-I, -II and -III demonstrated that these strains were still able to produce ethanol and benzyl alcohol and suggested that other oxidoreductases were catalysing these reductive biotransformations. Other authors [50, 51] have demonstrated the presence of ADH-IV, activated by Zn, in low amounts in *S. cerevisiae* and it is possible that this enzyme may play some part in ethanol and benzyl alcohol production.

Using yeast strain selection, immobilised cell technology and cyclodextrin addition, previous studies have reported maximal L-PAC concentrations of 10–$12\ \text{g}\,\text{l}^{-1}$. Recent studies by our group, however, have demonstrated with the control of yeast activity and metabolism (via control of the respiratory quotient,

Table 3. Comparison of kinetic evaluations for various methods of L-PAC production

Process	L-PAC $(g\,l^{-1})$	Biotrans. time (h)	Productivity $(g\,l^{-1}\,h^{-1})$	Yield (%) theoretical based on benzaldehyde	Ref.
(1) Batch and fed-batch processes					
Free cells	12.4	17	0.73	57	[21]
	22.4	14	1.6	65	[35]
Free cells (cyclodextrins)	12				[24]
Immobilised cells	9.9	3	3.3	60	[23]
	10	24	0.42	59	[22]
	15	22	0.7	58	[39]
Free PDC	28.6	8	3.6	95	[41]
Immobilised PDC	27.1	12	2.3	93	[49]
(2) Continuous processes					
Immobilised cells	4	–	0.6	45	[39]
Immobilised cells (semi-continuous)	4.5	–	0.4–0.8	57	[22]
Three-stage system (free cells)	10.6	–	0.44	56	[53]

RQ), that L-PAC concentrations up to $22\,g\,l^{-1}$ can be achieved in defined media in the normal biotransformation time. This has resulted from accumulation of pyruvic acid, induction of PDC and controlled feeding of benzaldehyde. It is interesting to note that other authors [52] using defined media and a *Candida* strain have also achieved relatively high levels of pyruvic acid by manipulation of PDC activity. Furthermore, the use of defined media also offers the advantages of easier solvent-extraction of L-PAC and, fewer pollution problems compared to the use of industrial substrates such as molasses.

From Table 3 it is evident that continuous biotransformation processes do not offer any advantages when compared to a fed-batch process under optimal control. This results from the highly toxic nature of the substrate benzaldehyde which considerably reduces cell viability and PDC activity in a continuous process. Rohner et al [36, 37] demonstrated that a chemostat culture was the most suitable for a biotransformation which involved the stereo-specific reduction of acetoacetic acid esters; however, in this case the inhibition effects of substrate/products on *S. cerevisiae* were much lower.

The use of purified PDC for the biotransformation demonstrated that the PDC could maintain its activity at benzaldehyde concentrations greater than $200\,mM$ $(21\,g\,l^{-1})$ and give L-PAC concentrations up to $28\,g\,l^{-1}$. Very high conversion efficiencies were achieved (greater than 95% theoretical based on benzaldehyde added) as no significant aromatic by-products were produced. However, with the purified PDC process it was necessary to add pyruvic acid with a mole ratio of pyruvate : benzaldehyde greater than 1.0, due to formation of the by-products acetaldehyde and acetoin. Mass balances on both the benzaldehyde and the pyruvate closed to within 2% indicating the accuracy of the analytical procedures and the identification of all major products in the biotransformation process.

In an overall assessment of the L-PAC process, it is clear that significant enhancements can be achieved by either improved process control of the yeast biotransformation or by the development of an enzymatic process based on purified PDC. The choice will ultimately be based on economic criteria; however, it may also be influenced by the development of genetically-engineered strains of yeasts with enhanced PDC and reduced oxidoreductase activities, resulting in higher L-PAC yields and productivities.

8 References

1. Creuger W, Creuger A (1988) In: Biotechnology: A Textbook of Industrial Microbiology, 2nd Edn. Science Tech. Publ., Madison, p.286
2. Feber K (1993) Biotransformations in Organic Chemistry. Springer New York Berlin Heidelberg
3. Klibanov AM (1986) Chemtech 16: 354
4. Halling PJ (1987) In: Moody GW, Baker PB (eds) Bioreactors and Biotransformations. Elsevier New York, p. 189
5. Aldercreutz P, Mattiasson B (1987) Biocatalysis 1(2): 99
6. Yamane T (1988) Biocatalysis 2(1): 1
7. Fukui S, Ahmed SA, Omata T, Tanaka A (1980) Eur J Appl Microbiol Biotechnol 10: 289
8. Zaks A, Klibanov AM (1988) J Biol Chem 263: 8017
9. Margolin AL, Klibanov AM (1987) J Am Chem Soc 109: 3802
10. Hedström G, Slotte JP, Bachlund M, Molander O, Rosenholm JB (1992) Biocatalysis 6: 282
11. Nakamura K, Chi YM, Yamada Y, Yano T (1986) Chem Eng Comm 45: 207
12. Kamat S, Barrera J, Beckman EJ, Russell A (1992) Biotech Bioeng 40: 158
13. Stinson SC (1993) Chem & Eng News, Sept. 27, 38
14. Hu S-Y (1969) Econ Botany 23: 346
15. Sørensen GG, Spenser ID (1988) J Am Chem Soc 110: 3714
16. Sørensen GG, Spenser ID (1989) Can J Chem 67: 998
17. Astrup A, Breum L, Toubro S, Hein P, Quaade F (1992) Int J Obesity 16(4): 269
18. Voets JP, Vandamme EJ, Vlerick C (1973) Z Allg Mikrobiol 13: 355
19. Netrval J, Vojtisek V (1982) Eur J Appl Microbiol Biotechnol 16: 35
20. Vojtisek V, Netrval J (1982) Folia Microbiol 27: 173
21. Culic J, Ulbrecht S, Vojtisek V, Vodansky M (1984) Method of cost reduction in the production of D-(-)-phenyl-1-hydroxy-2-propane for the production of L-(-) ephedrine. Czech Patent No. 22941
22. Mahmoud WM, El-Sayed AH, Coughlin RW (1990) Biotech Bioeng 36: 55
23. Seely RJ, Heefner DL, Hageman RW, Yarus MJ, Sullivan SA (1989) Process for producing L-PAC. An immobilised cell mass for use in the process and method for preparing cell mass. US Patent No. PCT/US89/64421
24. Mahmoud WM, El-Sayed AH, Coughlin RW (1990) Biotech Bioeng 36: 256
25. Seely RJ, Hageman RV, Yarus MJ, Sullivan SA (1989) Process for making L-PAC. Microorganisms for use in the process and method for preparing the microorganism. US Patent No. PCT/US 89/04423
26. Long A, Ward OP (1989) J Ind Microbiol 4: 49
27. Long A, James P, Ward OP (1989) Biotech Bioeng 33: 657
28. Long A, Ward OP (1989) Biotech Bioeng 34: 933

29. Nikolova P, Ward OP (1991) Biotech Bioeng 38: 493
30. Nikolova P, Ward OP (1992) Biotech Bioeng 39: 870
31. Nikolova P, Ward OP (1994) Biocatalysis 9: 329
32. Gupta KG, Singh J, Sahni G, Dhawan S (1979) Biotech Bioeng 21: 1085
33. Agarwal SC, Basu SK, Vara VC, Mason JR, Pirt SJ (1987) Biotech Bioeng 29: 783
34. Tripathi CKM, Basu SK, Vara VC, Mason JR, Pirt SJ (1988) Biotech Letts 10: 635
35. Wang B, Shin HS, Rogers PL (1994) In (Teo WK, Yap MGS, Oh SWK, eds) Better living through Innovative Biochemical Engineering. Continental Press, Singapore, p.249
36. Rohner M, Locher G, Sonnleitner B, Fiechter A (1988) J Biotechnol 9: 11
37. Rohner M, Münch T, Sonnleitner B, Fiechter A (1990) Biocatalysis 3: 37
38. Mahmoud WM, El-Sayed AH, Coughlin RW (1990) Biotech Bioeng 36: 47
39. Shin HS, Rogers PL (1995) Appl Microbiol Biotechnol 44: 7
40. Bringer-Meyer S, Sahm H (1988) Biocatalysis 1: 321
41. Shin HS, Rogers PL (1996) Biotech Bioeng 49: 52
42. Ullrich J, Donner I (1970) Hoppe-Seyler's Z Physiol Chem 351: 1030
43. Bioteux A, Hess B (1970) FEBS Letters 9(5) : 293
44. Lehmann H, Fisher G, Hübner G, Kohnert KD, Schellenberger A (1973) Eur J Biochem 32: 83
45. Urk H, Schipper D, Breedveld GJ, Paul RM, Scheffers WA, van Dijken JP (1989) Biochim Biophys Acta 992: 78
46. Scopes RK (1989) In: van Unden N (ed), Alcohol toxicity in yeast and bacteria, CRC Press Inc, p. 89
47. Jencks WP (1975) In: Meister A (ed), Advances in Enzymology and Related Areas of Molecular Biology, Wiley, New York, Vol. 43, p. 219
48. Perry RH (1984) Chemical Engineering Handbook, 6th Edn, McGraw-Hill International Edition, 3–27
49. Shin HS, Rogers PL (1996) Biotech Bioeng 49: 429
50. Buisson D, El-Baba S, Azered R (1986) Tetrahedron Lett 27: 4453
51. Drewke C, Ciriacy M (1988) Biochim Biophys Acta 950: 54
52. Besnainou B, Giani D, Sahut C (1990) Method for producing pyruvic acid by fermentation. US Patent No. 4,918,013
53. Wang B (1993) Kinetic study of fed-batch and continuous bioconversion processes for L-phenylacetylcarbinol (L-PAC) production by the yeast *Candida utilis*. PhD Thesis, University of New South Wales, Sydney, Australia

Inclusion Bodies and Purification of Proteins in Biologically Active Forms

Asok Mukhopadhyay

National Institute of Immunology, Aruna Asaf Ali Marg, New Delhi-110067, India

List of Symbols and Abbreviations 62
1 Introduction 64
2 Inclusion Bodies 65
  2.1 Mechanism of Inclusion Body Formation 65
  2.2 Physicochemical Characteristics of Inclusion Bodies 67
    2.2.1 Particle Size and Shape 67
    2.2.2 Density of the Particles 68
    2.2.3 Solubility of Inclusion Bodies 68
3 In Vivo Folding of Protein 69
4 Kinetics of Protein Folding 71
  4.1 In Vivo vs in Vitro Folding 72
  4.2 Proline Isomerization 72
  4.3 In Vitro Formation of Disulfide Bonds 73
    4.3.1 Chemistry of the Disulfide Bond 74
    4.3.2 Kinetics and Equilibrium of Thio-Disulfide Exchange 74
    4.3.3 Importance of Thiol Reagent 77
  4.4 Protein Aggregation In Vitro 77
5 Selection of Purification Protocol 81
  5.1 Isolation of Inclusion Bodies 83
  5.2 Purification of Protein in the Denatured State 87
  5.3 Refolding of Proteins into Bioactive Conformation 88
    5.3.1 Refolding by Dialysis, Diafiltration and Gel Filtration 88
    5.3.2 Refolding by Dilution 89
    5.3.3 Additive-Assisted Refolding 91
    5.3.4 Chemical Modification 92
    5.3.5 Refolding in Reversed Micelles 92
  5.4 Polishing of the Bioactive Molecules 93
6 Process Development and Scale Up 95
7 Removal of Extra Methionine from N-Terminus of Protein 95
8 Characterization of Recombinant Products 96
9 Purity and Safety 99
10 Case Studies 102
  10.1 Tissue Plasminogen Activator 102
  10.2 Insulin 102
11 Conclusions 103
12 References 105

Even though recombinant DNA technology has made possible the production of valuable therapeutic proteins, its accumulation in the host cell as inclusion body poses serious problems in the recovery of functionally active proteins. In the last twenty years, alternative techniques have been evolved to purify biologically active proteins from inclusion bodies. Most of these remain only as inventions and very few are commercially exploited. This review summarizes the developments in

Advances in Biochemical Engineering/
Biotechnology, Vol. 56
Managing Editor: Th. Scheper
© Springer-Verlag Berlin Heidelberg 1997

isolation, refolding and purification of proteins from inclusion bodies that could be used for vaccine and non-vaccine applications. The second section involves a discussion on inclusion bodies, how they are formed, and their physicochemical properties. In vivo protein folding in *Escherichia coli* and kinetics of in vitro protein folding are the subjects of the third and fourth sections respectively. The next section covers the recovery of bioactive protein from inclusion bodies: it includes isolation of inclusion body from host cell debris, purification in denatured state, alternate refolding techniques, and final purification of active molecules. Since purity and safety are two important issues in therapeutic grade proteins, the following three sections are devoted to immunological and biological characterization of biomolecules, nature, and type of impurities normally encountered, and their detection. Lastly, two case studies are discussed to demonstrate the sequence of process steps involved.

List of Symbols and Abbreviations

kDa	kiloDalton
k_m	rate constant of forward reaction for protein mixed disulfide
k_{-m}	rate constant of backward reaction
k_i	rate constant of forward reaction for protein disulfide
k_{-i}	rate constant of backward reaction
K_M	equilibrium constant of the formation of protein mixed disulfide
K_I	equilibrium constant of the formation of protein disulfide
K_O	overall equilibrium constant of the formation of protein disulfide
P_R	reduced protein
P_{DS}	oxidized (disulfide) protein
P_T	total protein
P_{MDS}	mixed disulfide protein
o	number of Arg residues
p	number of Lys residues
q	total number of Arg and Lys residues
r	number of Asp residues
s	number of Glu residues
t	total number of Asp and Glu residues
ACA	approximate charge average
bGH	bovine growth hormone
βhCG	β-subunit of human chorionic gonadotropin
DEAE	diethyl aminoethyl
DTT	dithiothreitol
EU	endotoxin unit
GSH	glutathione reduced
GSSG	glutathione oxidized
GuHCl	guanidine hydrochloride
hGH	human growth hormone
hG-CSF	human granulocyte colony stimulating factor
IEF	isoelectric focusing
IFN	interferon

IL	interleukin
GM-CSF	granulocyte macrophage colony stimulating factor
LAL	limulus amoebocyte lysate
PDI	protein disulfide isomerase
PEI	polyethyleneimine
PPI	prolyl-peptidyl *cis-trans* isomerase
QAE	diethyl [2-hydroxypropyl] ammonioethyl
RIA	radioimmuno assay
RNaseA	ribonuclease A
RSH	reductant
RSSR	oxidant
t-PA	tissue plasminogen activator

1 Introduction

Developments in recombinant DNA technology over the past decade have made possible the production of many scarce human therapeutic proteins in quantities unavailable before. Many of these proteins in the native form are non-glycosylated. Again, non-glycosylated protein has shown distinct therapeutic advantages over its native glycosylated form [1]. Therefore, large numbers of therapeutically important proteins are now being produced in E. coli, the workhorse of recombinant DNA technology, many of which are in clinical use, while some others are undergoing clinical evaluation.

The majority of the expressed heterologous proteins accumulate in the cytoplasm of E. coli due to the absence of a proper secretory mechanism. Purification of these proteins into therapeutic grade is tricky. Attempts have been made to secrete these proteins into the periplasm with the help of a signal peptide of a membrane or secretory protein. So far, such attempts have met with only limited success due to the complex nature of the heterologous proteins and their secretion mechanism. Therefore, many therapeutic proteins are still produced as inclusion bodies expressed in the cytoplasm. In the near future, a good number of these proteins will continue to be produced in a similar system. Cytokines thus produced command a present market value of over $750 million. This is likely to grow almost ten-fold by the year 2000 [2]. Production of hormones and cytokines via the route of inclusion bodies encompasses not only the challenge of understanding protein folding, but now also the challenge of large scale production of therapeutic grade products in the most competitive manner.

The formation of inclusion bodies offers several advantages for the production of heterologous proteins. These proteins may be unstable in the cytoplasm of a bacterium due to proteolysis (e.g., insulin chains A and B) and may be toxic to the host cell in the native conformation (e.g., immunotoxins). Inclusion bodies of heterologous proteins generally represent 5–40% of the total cellular proteins [3,4]. Because of their high concentration and aggregated form, inclusion bodies can be isolated from the cellular proteins in highly purified form (50–70% of total Coomassie blue stainable protein). This obviously enables cost-efficient downstream processing [5–7]. However, the main problem, namely a technical breakthrough for an efficient route for the production of biologically active proteins from the aggregated and denatured inclusions useful for therapeutic purposes, remains. In order to facilitate the purification of biologically active proteins, it is necessary to understand why inclusion bodies are formed, how proteins from inclusion bodies are folded in vitro, and what are the alternate routes for the same. For therapeutic grade protein, safety and purity are the two most important considerations. It is crucial to understand how these affect purification schemes. The review strives to provide an update on current practices in the area of the purification of functional proteins from the inclusion bodies intended to be used for clinical applications.

2 Inclusion Bodies

Inclusion bodies are aggregated, extremely dense structures of protein, produced mostly in the cytoplasm of *E. coli*. These inclusion proteins have partial secondary structure. Inclusion bodies can be as large as a normal bacterial cell. They change the light-scattering properties of the cultures [8]. The larger inclusions are visible under an optical microscope as particles reflecting light, and termed "refractile bodies". They are morphologically different from any other intracellular structures found in the cytoplasm of prokaryotes. Figure 1 presents the distinct structure of a fusion protein β-galactosidase/insulin chain-A, expressed in *E. coli* as inclusion bodies. A list of approved therapeutic proteins and some of those in clinical trials are shown in Table 1.

2.1 Mechanism of Inclusion Body Formation

Despite extensive studies on inclusion bodies over the last 15 years, the mechanism of their formation is still obscure. Previous reviews are mainly concerned with the kinetics of aggregation, solubility of recombinant proteins, and their expression in soluble form [12–14]. This review describes a few important factors believed to be responsible for the formation of inclusion bodies.

Many eukaryotic genes have been cloned and expressed in *E. coli*. A great majority of these proteins are found to be produced in the cytoplasm as inclusion bodies. The general view has been that inclusion bodies are produced

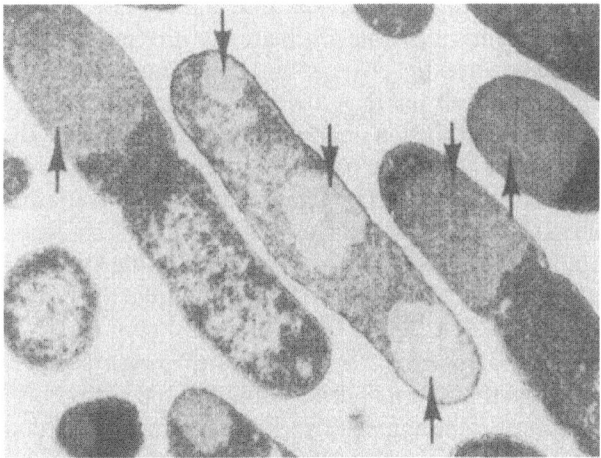

Fig. 1. Transmission electron micrographs of *E. coli* cells producing fusion protein of βgal-insulin chain A (x17,500). *Arrow* indicates distinct inclusion bodies of fusion protein. From William et al. [9], reproduced with permission

Table 1. Therapeutic proteins obtained from recombinant *E. coli*

Protein	Mw (kDa)	Clinical treatment	FDA approved/Trials
Insulin	6	Diabetes	1982 (approved)
hGH	22	hGH deficiency in children	1985 (approved)
IFNα	18	Hairy cell leukemia, genital warts, AIDS-related Kaposi's sarcoma, non-A, non-B hepatitis	1986–1991 (approved)
IFNγ	34 (dimer)	Infection/chronic granulomatous disease	1990 (approved)
G-CSF	16	Induced neutropenia	1991 (approved)
GM-CSF	30	Infection related to bone marrow transplant	1991 (approved)
IL-2	17	Chemotherapy	1992 (approved)
IFNβ	20	Multiple sclerosis	Phase III

See [10, 11]

only in the cytoplasm of *E. coli*. Recently, it has become clear that inclusion bodies are produced in other host cells, like *Bacillus sp.* [15], *Saccharomyces sp.* [16], insect cells [17], and monkey cell line BSC-40 [18] as well. These observations thus indicate that the formation of an inclusion body is not host-specific.

The nature of the expressed proteins, the rate of their expression, and the level of expression exert a profound influence on the formation of inclusion bodies. High expression rate allows insufficient time for the nascent polypeptide chain to fold into the native conformation. This, combined with the high local concentration in the cytoplasm, leads to non-specific precipitation. The restriction enzyme *EcoRI* is present as a soluble dimeric protein at a low level of expression [19]. However, at higher expression rates, it produces tetrameric inactive inclusion bodies [20]. These observations suggest that a higher expression rate facilitates the formation of tetrameric or even higher forms of *EcoRI*, which ultimately precipitate. On the other hand, aspartase and cyanase are two homodimeric, soluble, stable proteins in *E. Coli* which are not precipitated even at a very high level (30%) of expression [21]. A combination of factors relating to the physiological state of the host cell and the growth conditions are reported to determine the formation of inclusion bodies. The majority of the prochymosin inclusion bodies are found to be produced in the late exponential to early stationary phases of growth [22]. The temperature of cultivation is another important variable in the formation of inclusion bodies. The expression level of human interferon-β (IFNβ) has been observed to be the same at 20 and 33 °C. However, the formation of the active form of IFNβ is reported to be eight times higher at 20 than at 33 °C [23]. A similar observation has been made for IFNγ, where 30–90% active form is obtained when the cultures grow at temperatures below 30 °C. This suggests that thermal denaturation of the native form of the protein may be responsible for the aggregation. It has been proposed that the proteins characterized by higher denaturation temperature and higher native state stability do not form inclusion bodies [12]. However, this theory of high denaturation temperature has been rendered null-and-void by the results of in vivo folding of bacteriophage P22 tailspike protein. P22 tailspike protein forms inclusion bodies at a temperature ranging from 35 to 40 °C, despite having an extremely high denaturation temperature (88 °C). Finally, it appears that the intermediate conformations of the polypeptide chains are sensitive to temperature [24].

The cytoplasm of *E. coli* is maintained at a reducing environment due to the presence of higher proportion of reduced glutathione [25]. The reducing environment in the cytoplasm prevents the formation of disulfide bonds. Thus, the proteins requiring disulfide bonds to assume their native conformation, are believed to be aggregated in the cytoplasm. Again, there are many exceptions to this process. Proteins without disulfide bond (e.g., *E. coli* sigma protein, IFNγ) or cysteine still accumulate in the cytoplasm as inclusion bodies [7, 26]. Furthermore, though the periplasm of *E. coli* is non-reducing in nature [27], protein secreted in it is found to be aggregated. Georgious et al. [28] have shown that over-expression of β-lactamase in the periplasm leads to the formation of inclusion bodies.

Heat shock/chaperone proteins are a family of unrelated classes of proteins that prevent protein aggregation in the cytoplasm and help in the transport and folding of proteins [29, 30]. Aggregation may occur when a heterologous protein expressed in *E. coli* is either unable to bind to the host cell chaperones, or binds very tightly to prevent dissociation. It is also possible that the concentrations of chaperone proteins within the cell are not commensurate with the amount of protein expressed at a given time. Thus, whereas in *E. coli* X 156, β-lactamase is expressed in the soluble form, it is aggregated in *E. coli* CAG 456 [12, 31]. *E. coli* CAG 456 is a mutant of htpR gene that controls the heat shock genes including lon (Cap R).

The foregoing reports on the formation of inclusion bodies tempted the author to conclude that inclusion bodies may be produced in the cytoplasm of any cell type, and are formed due to any or all of the following reasons: heterologous nature of the protein; protein expressed at a higher rate; relatively more hydrophobic protein which aggregates intermolecularly as a result of non-covalent association; and chaperones like helper proteins are either inadequately available or absent.

2.2 Physicochemical Characteristics of Inclusion Bodies

Information available on the physicochemical characteristics of inclusion bodies is very scant. The following observations have been found very useful in the isolation and subsequent solubilization of these inclusion bodies.

2.2.1 Particle Size and Shape

Since inclusion bodies are amorphous or paracrystalline in nature, they make the protein densely packed and impervious to proteolytic activity. The size and shape of the inclusion bodies in the cytoplasm have been observed to be different from those in the periplasm. Ultrastructure studies on purified inclusion bodies of β-lactamase localized in the cytoplasm reveals them to be of highly regular and cylindrical shape of average length of 1.5 μm [32]. The mean particle sizes of IFNγ and prochymosin are found to be 0.81 μm and 1.28 μm with a standard deviation of 0.17 μm and 0.46 μm respectively [33]. The size and shape of these particles could vary over a wide range. In the author's laboratory, inclusion particles of human growth hormone (hGH) and β-subunit of human chorionic gonadotropin (βhCG) hormone have been observed in highly irregular forms. The sizes of the particles in the buffer were found to be enlarged due to hydrophobic interaction

2.2.2 Density of the Particles

The compact nature of inclusion bodies make it easier to recover them from cell debris by centrifugation. Information on the size and density can help to determine the probable time and centrifugal force (x g) required to recover these particles from the cell homogenate. The effective buoyant density of the larger particles is high, so they are easily isolated from the lighter particles. The buoyant densities of IFNγ and prochymosin inclusion bodies is cesium chloride solution are reported to be 1340 and 1240 $kg\,m^{-3}$ respectively [33].

2.2.3 Solubility of Inclusion Bodies

Aggregates of protein are stabilized by the intermolecular hydrophobic interactions, involving specific hydrophobic faces in the peptide chains. Solubilization of these aggregates requires disruption of the forces that hold them together, by using chaotropic agents or detergents. In vitro solubility of a protein inclusion body is dependent on two parameters, namely hydrophobicity and approximate charge average.

2.2.3.1 Hydrophobicity

The protein contains many amino acid residues having hydrophobic aromatic and aliphatic side chains. In a properly folded protein, many of these hydrophobic groups are buried inside, thus minimizing the contact area with the water molecule. Rose et al. [34] have grouped amino acids into three categories based on linear relationships of their average area buried upon folding and their solvent accessible surface area. These are (i) hydrophobic (fully buried: Ala, Cys, Val, Ile, Leu, Met, Phe, Trp), (ii) moderately polar (80% buried: Ser, Thr, His, Tyr), and (iii) very polar (~60% buried: Asp, Asn, Glu, Gly, Arg, Pro). The hydrophobicity of a protein is determined by analysis of the hydropathy indices of its constituent amino acids. The indices are summed and divided by the number of amino acid residues to yield a "gravy score", or a measure of relative hydrophobicity of the protein [35, 36]. The solubility of a protein reduces with the increase in hydrophobicity.

2.2.3.2 Approximate Charge Average

Proteins are polyelectrolytes carrying both positive and negative charges. At low salt concentrations, solubility of a protein can be predicted by the Debye-Hückel relation [37]. According to this relation, log of protein solubility is proportional to the square of the net protein charge. Thus, protein solubility increases with increasing net charge, which may be positive or negative. At neutral pH, the net charge of a protein is very close to the difference between the number of basic and acidic amino acid residues [38], as shown in Eq. (1). Wilkinson and Harrison [38] observed a good correlation between approximate charge average and inclusion body formation in the cytoplasm of a bacterial cell.

Net charge = Σ Basic amino acids - Σ Acidic amino acids

$$= \Sigma[(Arg)_0 + (Lys)_p]_q - \Sigma[(Asp)_r + (Glu)_s]_t$$

$$\textbf{Approximate charge average (ACA)} = \frac{\textbf{Net charge}}{\Sigma \textbf{Amino acids}} \qquad \textbf{(1)}$$

3 In Vivo Folding of Protein

In vivo protein folding in prokaryotic and eukaryotic cells occurs by overlapping of sequential processes, such as translation, translocation, proteolytic processing, post-translational modifications, and assembly. The nascent polypeptide chain has to go through these steps in order to become a functional protein. Details of these processing steps, elaborated elsewhere [39, 40], are beyond the scope of this review. However, a brief overview of the mechanism of export and oxidative folding of proteins in E. coli is discussed in this section. This will help the reader comprehend the basics of the mechanism of in vivo folding.

Proteins are synthesized on the ribosome present in the cytosol by the sequential addition of amino acids to the N-terminus. The rate of addition of each amino acid to the growing polypeptide is substantially high. Within 10 to 1000 s most of the polypeptides, whatever their size, are synthesized [39]. In the case of small and simple proteins (without disulfide bond), folding occurs cotranslationally in the cytosol. Most of the eukaryotic proteins expressed in E. coli require correct disulfide pairing to become biologically-active molecules.

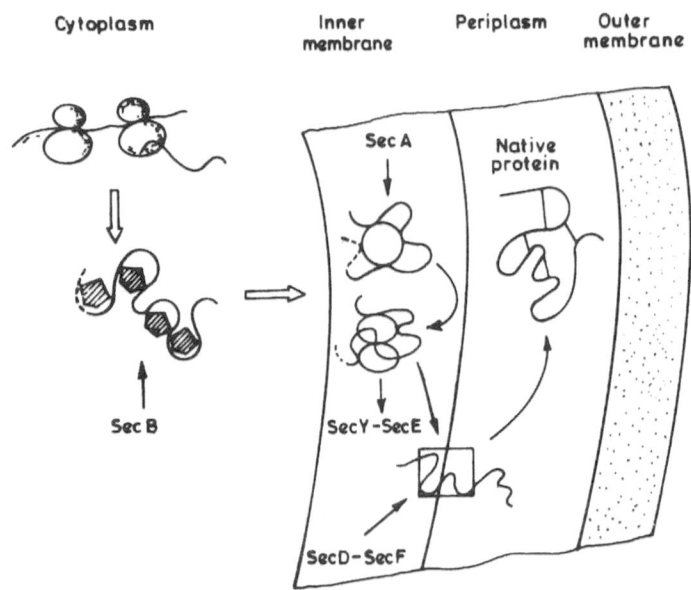

Fig. 2. In vivo synthesis of protein, its translocation across the membrane and oxidative folding in the periplasmic space of *E. coli*

Oxidative folding of these proteins in the cytosol of *E. coli* is prevented [41] due to the very high reducing environment (GSH/GSSG = 100–400). On the other hand, periplasm extends a congenial environment for folding due to its non-reducing nature, and the presence of high energy protein disulfide reagents (DsbA) and prolyl-peptidyl *cis-trans* isomerase (PPI). Thus, considerable interest has been shown in expressing heterologous eukaryotic proteins in the periplasm.

E. coli contains both an outer membrane and a cytoplasmic inner membrane. Periplasm is the region between these two membranes. To fold the protein into a bioactive conformation, it has to be translocated from the site of synthesis to the periplasmic space (Fig. 2). A number of heterologous proteins have been secreted post-translationally into the periplasm, or in the extracellular milieu with the help of an N-terminus extension, called signal/leader peptide. These signal sequences (both prokaryotic and eukaryotic in origin) are typically 15–30 amino acid residues long, containing a hydrophobic segment that is preceeded by a few positively charged residues [42]. Signal peptides generally used in the secretion of the protein in *E. coli* are periplasmic proteins such as alkaline phosphatase, or proteins of the outer membrane like OmpA, OmpF, etc. Evidence suggests that transport across the cytoplasmic membrane requires a translocation-competent protein conformation which is in soluble form, but not completely folded [43]. Translocation-competent state is equivalent to the "molten globule" state of the protein which is highly dynamic, energetically closer to the unfolded state, and possesses some secondary structure [44]. The function of the signal peptide has been shown to provide direction, and to slow down the folding of the protein to be transported into the periplasm. A number of studies have shown that in *E. coli* the cytoplasmic chaperone SecB preferentially interacts (non-covalently) with those polypeptides (preproteins, the proteins containing the signal peptide) destined to be secreted. SecB chaperone stabilizes the preprotein in the form of a translocation-competent state. In vitro translocation of pre-maltose binding protein (MBP) into plasma membrane is found to be dependent on the addition of purified SecB [45]. GroEL (hsp 60) is another class of chaperone that binds to and arrests the folding of the precursor form (pre-β-lactamase) of the secreted protein [46]. Similarly, another cytosolic protein, "trigger protein", has been found capable of maintaining a translocation competent state of the precursor proteins [47]. The energy for the translocation process is provided by the membrane potential and by ATP hydrolysis [48, 49].

Preprotein is transferred from SecB to the peripheral membrane chaperone SecA by direct interaction. Both the signal peptide and the matured parts of the protein are bound with SecA. SecA containing preprotein then interacts with the integral membrane SecY-SecE translocase complex, and ATPase activity results in the dissociation of the protein from SecA. The protein is ultimately transported from the cytosol and through the inner membrane of *E. coli* by the concerted efforts of SecB, SecA, and SecY-SecE systems. The post-translocational step of the secretory protein is mediated by a trans-membrane protein complex termed SecD-SecF, largely exposed to the periplasm. Cleavage of the signal peptide occurs by the action of membrane-bound enzymes. The SecD-SecF system prevents any side reactions and is also involved in the release of the protein, which completes its folding in the periplasm [48, 50, 51].

The oxidative folding of translocated protein in the periplasm is catalyzed by a significantly stronger oxidant and high energy disulfide bond containing protein, known as DsbA. It is functionally equivalent to protein disulfide isomerase (PDI) of eukaryotic systems which catalyzes correct disulfide pairing of periplasmic *E. coli* proteins in vivo and in vitro [52, 53]. DsbA has been found to be involved in the production of heterologous eukaryotic proteins, such as antibody fragments [54] and serine proteases [55] in the periplasm of *E. coli*. The rate-limiting step in protein folding is *cis-trans* isomerization of the X-Pro peptide bonds (X: any amino acid). PPI, another periplasmic protein, is found to accelerate folding by catalyzing this rate-limiting isomerization of peptide bonds of translocated protein [25, 56].

4 Kinetics of protein folding

The term denaturation refers to any change from the biologically active conformation (native state), including irreversible covalent modification and/or aggregation. Proteins unfold in strong denaturant (e.g., 6 mol/l GuHCl or 8 mol/l urea), and attain the average hydrodynamic properties similar to the random coil polypeptide [57]. The interactions that stabilize folded proteins (such as hydrogen bonding and hydrophobic interactions) are drastically reduced in the denatured protein. The polypeptides tend to form small α-helical conformation on the basis of a favorable local sequence of amino acids [58]. The refolded proteins consist of native-like secondary structure, which interact with each other and are further stabilized [59, 60]. Disulfide bonds stabilize protein by decreasing the conformational entropy of the unfolded state [61]. A number of theoretical models for protein folding have been discussed in the literature [62, 63]; these are beyond the scope of this review. The present section deals with some of the limiting steps controlling the kinetics of folding in vitro.

4.1 In Vivo vs In Vitro Folding

In vivo protein folding generally yields 100% native conformation. However, the protein when refolded in vitro from the denatured state may yield 0–100% native conformation, depending on the type of the protein and the refolding conditions employed. In vivo folding of proteins in mammalian systems is believed to occur co-translationally. The final folding is determined by the kinetics and thermodynamic state of the protein [64]. On the other hand, in vitro protein folding is mainly decided by the thermodynamic constraints. Accumulated evidence suggests that there are three important events involved in the rate-limiting steps of in vivo protein folding. These are isomerization of prolyl-peptide bonds, formation of disulfide bonds, and monomer association catalyzed by specific helper proteins [65, 66]. Generally, in vitro folding is conducted without these helper proteins, and thus is a less efficient process.

4.2 Proline Isomerization

In the denatured state, prolyl-peptide bonds (amide linkages) freely isomerize and attain an equilibrium mixture of prolyl isomers as shown in Fig. 3. The isomers differ by a 180° rotation around the carbonyl carbon to amide nitrogen (C-N) bond. At equilibrium, the contents of cis and trans isomers in the unfolded polypeptide are governed by the local amino acid sequence and its composition is close to 30 and 70% respectively [67]. In native proteins cis isomers are not uncommon, but are less prevalent (7%) than the trans isomers [25]. For correct folding to occur, each prolyl-peptide bond in the polypeptide chain must be similar to the native format. The polypeptide chains with the correct set of isomers (like native) usually refold faster than the incorrect isomers. The polypeptide chain with one or more incorrect prolyl-peptide isomers must first isomerize into the native format before initiation of folding. As a result, the overall refolding rate is found to be relatively slower [67]. The folding of protein

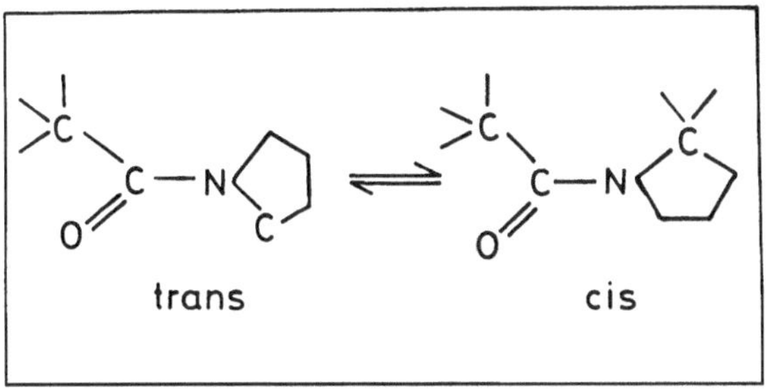

Fig. 3. *Cis-trans* isomerization of prolyl-peptidyl bond (carbonyl carbon to amide nitrogen)

Table 2. Number of theoretically possible combinations of disulfide pairing from disulfide bonds of a protein

Protein	Disulfide bonds	Combinations
IL-2	1	1
bGH/hGH	2	3
Insulin	3	15
RNaseA	4	105
βhCG	6	10395
t-PA	17	6.33×10^{18}

containing high proline is therefore unusually slow as it increases the chance of the non-native isomeric form. The isomerization of non-native form of the proline into native form occurs in an unfolded state devoid of the involvement of secondary or tertiary structures [68]. The process of isomerization is accelerated by PPI, an enzyme discovered by Fischer et al. [69]. In eukaryotes, the enzyme is located in the endoplasmic reticulum (ER). PPI is found to catalyze folding of the immunoglobulin light chain and ribonucleaseA (RNase A) in vitro [70]. Both PPI and PDI carry out catalysis in a concerted fashion during in vitro and in vivo folding, thus improving the catalytic efficiency of PDI [71].

4.3 In Vitro Formation of Disulfide Bonds

Disulfide bond formation with concomitant folding complicates the in vitro regeneration of native proteins from their reduced and denatured state. The complexity further increases with the increase in the number of half-cystines. Table 2 shows the number of ways in which two half-cystines combine to form a disulfide bond [72, 73]. For each protein only one of several theoretically possible combinations (Table 2) represent the native pairing. Most of these conformations are thermodynamically unstable and so do not exist in nature. Under optimal conditions of in vitro folding, the equilibrium is favored towards native pairing, and thus the majority of the polypeptide chains attain their native conformation. Before discussing oxidative refolding in detail, it is prudent to know the mechanism of the disulfide bond formation.

4.3.1 Chemistry of the Disulfide Bond

The formation of disulfide bonds from thiols of a protein is a two-electron oxidation reaction requiring the presence of an appropriate electron acceptor (oxidant) as shown in Eq. (2). Protein thiol oxidation to the disulfide is carried out by low molecular weight disulfide reagents, such as oxidized glutathione or cystamine. The cytoplasm of prokaryotes and eukaryotes is highly reducing in nature, so the equilibrium constants for thiol-disulfide exchange processes are unfavorable for the formation of disulfide bonds [41]. In eukaryotes, the formation of disulfide bonds is more favorable in the secretory pathway because ER is significantly oxidizing due to the low ratio of GSH/GSSG, and PDI resides in the lumen of the ER [41, 74]. In a gram-negative bacterium, the disulfide bond formation occurs in the periplasm with the help of DsbA [52].

$$P\overset{\displaystyle\diagup SH}{\diagdown SH} \rightarrow P\overset{\displaystyle\diagup S}{\diagdown S}| \;+\; 2H^+ + 2e \qquad\qquad (2)$$

Thiol-disulfide exchange occurs via direct nucleophilic attack of the thiolate anion ($-S^-$) on one of the sulfurs of the oxidant, forming a transient intermediate. As a result of nucleophilic substitution, a sulfur atom is removed and a mixed disulfide intermediate is formed. By an intramolecular displacement, disulfide bond formation is completed in the protein, as shown in Eq. (3). The formation of a disulfide bond is favored under the following two circumstances: (a) at pH of protein solution between 8–9, to favor the formation of the thiolate anions ($-S^-$), and (b) when the two Cys-residues are close to each other by few Å [75].

$$P\overset{\displaystyle\diagup S^- H^+}{\diagdown SH} + \overset{\displaystyle S-S}{\underset{R\ \ R}{|\ \ |}} \leftrightarrow \left[P\overset{\displaystyle\diagup S - S \overset{\delta^-}{\frown} S}{\diagdown SH\ \ R\ \ R} \right]^{\#} \leftrightarrow P\overset{\displaystyle\diagup S - SR}{\diagdown SH} + HSR \leftrightarrow P\overset{\displaystyle\diagup S}{\diagdown S}| + HSR \quad (3)$$

4.3.2 Kinetics and Equilibrium of Thiol-Disulfide Exchange

As mentioned earlier, the protein disulfide bonds are formed or broken in vitro in two sequential thiol-disulfide exchange reactions in a low molecular weight redox buffer. The reaction is reversible and involves the formation of a mixed disulfide intermediate (P_{MDS}) by an intermolecular reaction, followed by an intramolecular displacement to complete protein disulfide bond (P_{DS}) formation as shown in Eq. (4).

$$
\underset{(P_R)}{P\overset{SH}{\underset{SH}{\diagdown}}} + RSSR \xrightarrow[k_{-m} \; RSH]{\overset{K_M}{\underset{k_m}{}}} \underset{(P_{MDS})}{P\overset{S-SR}{\underset{SH}{\diagdown}}} \xrightarrow[k_{-i} \; RSH]{\overset{K_I}{\underset{k_i}{}}} \underset{(P_{DS})}{P\overset{S}{\underset{S}{\diagdown|}}} \qquad (4)
$$

The disulfide component of the redox buffer provides oxidizing equivalents, whereas the thiol component serves to reduce native or wrongly paired disulfide bonds. The disulfide and thiol reagents normally used in the redox buffer are listed in Table 3. The rate of formation of protein disulfide bonds depends on the conformation of the thiol group of the protein, nature of the disulfide reagent employed, accessibility and reactivity of the thiolate anion, and the conformational changes required to make the disulfide bonds [75]. The equilibrium constants for the formation of P_{MDS} and P_{DS} are, by and large, dependent on the concentrations of thiol-disulfide reagents and the redox potential of the buffer [41]. Under conditions where P_{MDS} do not accumulate significantly, the overall equilibrium constant (K_O) for bimolecular reaction between P_R and RSSR to form P_{DS} and RSH is given by Eq. (5). The fraction of P_R that remains at equilibrium can be determined from the composition of the redox buffer, defined by $(RSH)^2 (RSSR)^{-1}$. At $(RSH)^2 (RSSR)^{-1} \geq K_O$, Eq. (5) reduces to $P_R = 0.5 \, P_T$, that means half of the protein is oxidized to disulfide. More protein disulfide (P_{DS}) should be expected, when $(RSH)^2(RSSR)^{-1} \leq K_O$. At equilibrium, the abundance of early folding intermediates of oxidation will depend on their relative thermodynamic stability and the composition of redox buffer in which the protein is folded [76, 77]. The protein molecules having the most stable disulfide bond will be present at the highest concentration. There could be two reasons for the inadequate formation of disulfides in protein folding, as follows.

$$
K_O = K_M K_I = \frac{k_m}{k_{-m}} \cdot \frac{k_i}{k_{-i}} = \frac{|P_{DS}| \, |RSH|^2}{|P_R| \, |RSSR|}
$$

$$
= \frac{|P_T - P_R|}{|P_R|} \cdot \frac{|RSH|^2}{|RSSR|}
$$

$$
= \left[\frac{P_T}{P_R} - 1 \right] \frac{|RSH|^2}{|RSSR|}
$$

$$
\text{or,} \quad \frac{|P_T|}{|P_R|} = \frac{K_O + |RSH|^2/|RSSR|}{|RSH|^2/|RSSR|} \qquad (5)
$$

Table 3. Different thiol and disulfide reagents

A. Thiol Reagents

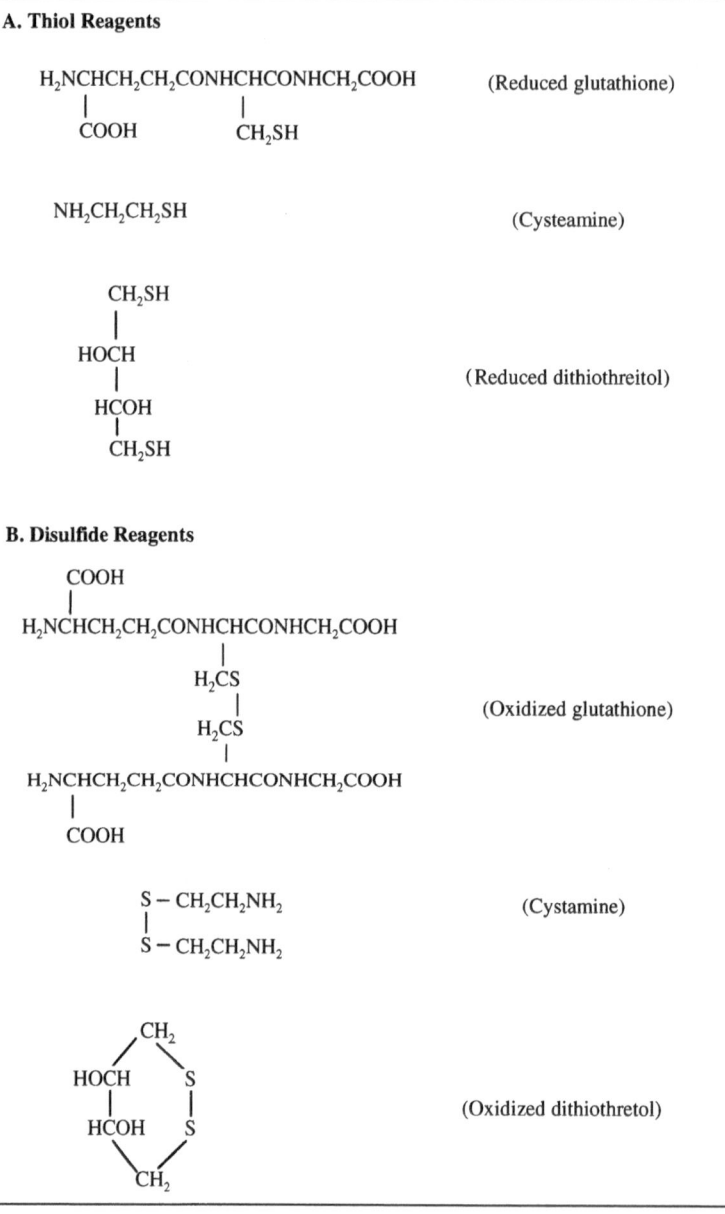

H₂NCHCH₂CH₂CONHCHCONHCH₂COOH (Reduced glutathione)

|

COOH CH₂SH

NH₂CH₂CH₂SH (Cysteamine)

CH₂SH
|
HOCH
|
HCOH
|
CH₂SH
(Reduced dithiothreitol)

B. Disulfide Reagents

COOH
|
H₂NCHCH₂CH₂CONHCHCONHCH₂COOH
|
H₂CS
|
H₂CS
|
H₂NCHCH₂CH₂CONHCHCONHCH₂COOH
|
COOH
(Oxidized glutathione)

S – CH₂CH₂NH₂
|
S – CH₂CH₂NH₂
(Cystamine)

CH₂
HOCH   S
| |
HCOH   S
CH₂
(Oxidized dithiothretol)

1. Conformational restriction preventing two half-cystines from forming a disulfide bond. In unfolded proteins without significant secondary structure, the cysteine residues (half-cystine) are far apart, and thus K_i is very small. In such a case, the rate-limiting step is intramolecular displacement. The rate of

intramolecular displacement to complete protein disulfide bonds is dependent on the number of intervening amino acids between two cysteines [78]. The overall equilibrium constant (K_O) for disulfide bond formation is reduced with the increase in the number of intervening amino acids. The accumulation of mixed disulfide (P_{MDS}) results due to conformational restrictions. In extreme cases, intramolecular disulfide formation is inhibited due to the formation of species where both protein thiols have been converted into mixed disulfides.

2. Inaccessibility and unreactivity of the protein thiols limiting the formation of mixed disulfides, and hence the protein disulfide bonds. Inaccessibility could be due to buried thiolate anion $(-S^-)$ in the folded protein. In the presence of weak denaturant, the polypeptide gets the proper environment for folding. The cys-residues are in close proximity in a partially folded protein, thus allowing for the formation of protein disulfide bonds very rapidly from the mixed disulfides.

4.3.3 Importance of Thiol Reagent

Oxidative refolding of reduced and denatured protein surprisingly requires significant amounts of reductant (RSH), normally ten times higher than the oxidant (RSSR). In the absence of RSH, oxidative folding is inhibited by the formation of non-productive intermediates and the protein is known to be kinetically trapped. Protein molecules are rescued from this trap by thiol-disulfide exchange reaction using suitable proportion of reduced-oxidized thiols.

Figure 4 shows a hypothetical folding pathway of a peptide having 12 half-cystines. The peptide is expected to be folded into native conformation (N) provided it follows the pathway leading to an intermediate (I). However, very often the protein is wrongly folded due to mis-pairing of disulfide bonds. Proteins are trapped into wrong conformations by (a) formation of stable non-native disulfide bonds as shown by the conformation (A), (b) formation of disulfide bonds exposed to the solvent prior to the formation of disulfide bonds at the core (B), (c) formation of mixed disulfides in the complementary half-cystines (C), and (d) a combination of (a) and (b), resulting in cysteine residues in the reduced state (D). The trapped molecules are released by intermolecular thiol-disulfide exchange using RSH. Thus, the efficiency of the oxidative refolding process is improved. Table 4 shows the amount of RSSR and RSH that has been used in the redox buffer for in vitro folding of different proteins.

4.4 Protein Aggregation In Vitro

How proteins aggregate in vitro has been the subject of much discussion for a long time. Aggregated proteins are solubilized in 6 mol/l GuHCl or in 8 mol/l urea. The denatured proteins are refolded at lower concentrations of denaturing

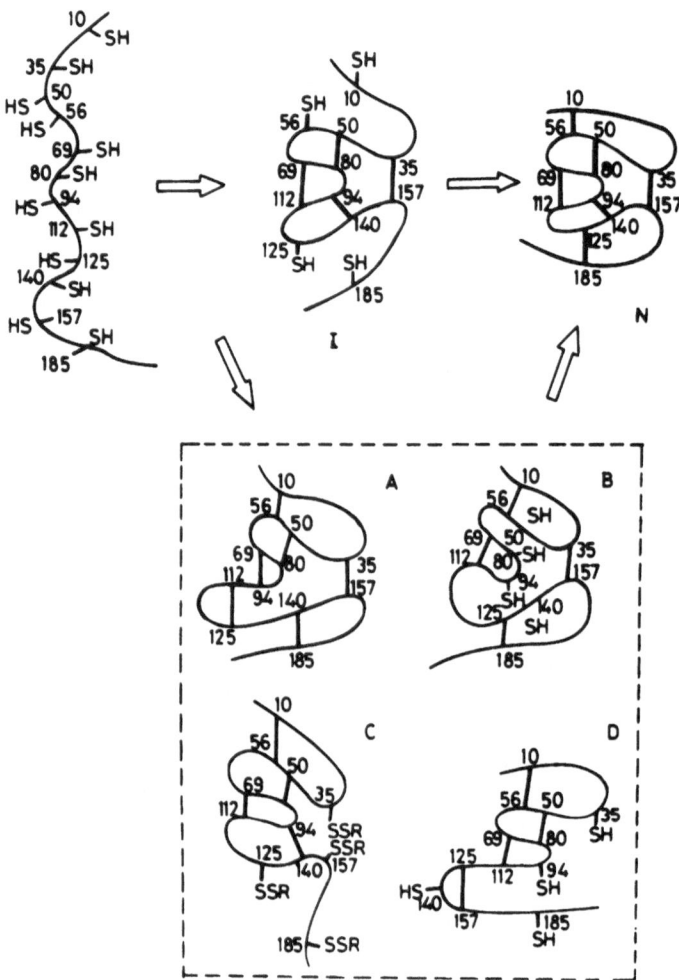

Fig. 4. Hypothetical folding pathway of a peptide containing 12 half-cystines, and kinetically trapped disulfide isomers in absence of thiol reagent. I: Intermediate of folding pathway (only one intermediate is shown); N: Native conformation; A: Non-native disulfide bonds; B: Isomer without disulfide bond in the core of the molecule; C: Isomer with mixed disulfides in the complementary half-cystines; D: Non-native disulfide bonds with free thiol groups (other than these four, there may be many disulfide isomers)

agents, where the protein structure is stabilized by the hydrogen bonding and hydrophobic interactions. Often, the protein is precipitated/aggregated when transferred from a higher to a lower denaturing buffer.

A high degree of aggregation of tryptophanase has been observed during refolding in 3 mol/l urea. It finally yields very little active enzyme in urea-free buffer [84]. Aggregation increases with protein concentration. Precursors of this aggregated protein are believed to be the intermediates of the protein folding pathway. The intermediates of the refolding pathway of porcine muscle lactate

Table 4. Composition of oxidant and reductant used in the thiol-disulfide exchange of different proteins

Protein	Oxidant (mmol/l)	Reductant (mmol/l)	Reductant/ Oxidant	Reference
Prorenin	0.1	1.0	10	[79]
Urokinase	1.0	10.0	10	[79]
RNaseA	0.2	2.0	10	[80]
Lysozyme	0.3–0.6	3.0–6.0	10	[81]
Fab-fragment	0.5	5.0	10	[82]
βhCG	2.5	5.0	2	[83]

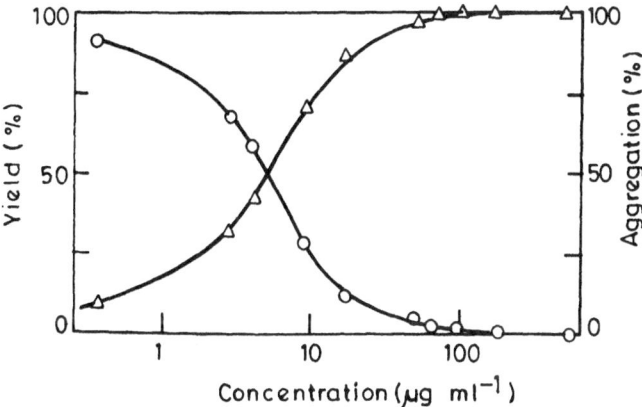

Fig. 5. Effect of initial protein concentration on refolding yield of LDH and its aggregation. (○): Yield of refolding; (△): Aggregation. From Zettlemeissl et al. [86], reproduced with permission

dehydrogenase (LDH) are partially folded monomeric form, with either end in native form, or aggregate via nonproductive pathway [85, 86]. Similar to tryptophanase, the yield of the active LDH is found to be sensitive to the initial concentration of protein in the folding reaction, as depicted in Fig. 5. The circular dichroism spectrum of the aggregated LDH shows the presence of partial secondary structure [86]. In the aggregated LDH, the monomeric forms are held by strong noncovalent forces, which is destabilized in 6 mol/l GuHCl or at low pH [86]. In a similar study, it has been found that the partially folded species of horse muscle phosphoglycerate kinase (PGK) act as precursors of aggregated protein [87]. The monomers are associated by specific interaction of hydrophobic β-sheets with the surface of α-helices of partially folded monomer. In bovine growth hormone (bGH) the putative amphiphilic helix comprising residues 110–127 amino acids has been found to play a critical role in the association processes. The aggregation is prevented when the peptide fragment

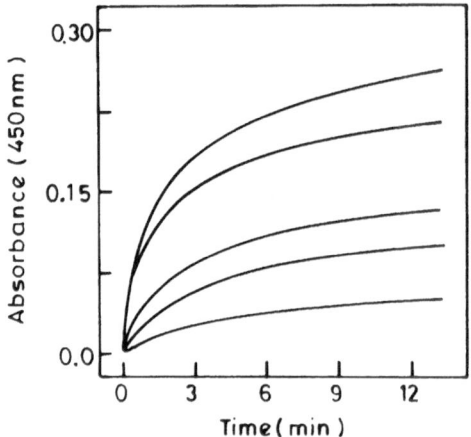

Fig. 6. Inhibition on the aggregation of bGH in presence of different amounts of peptide 96–133. Aggregation of protein at different times and peptide concentrations were measured by monitoring the absorbance at 450 nm. The *top curve* represents control absorbance in absence of peptide. The *bottom curves* (descending order from the top curve) were obtained by using peptide fragment at a 1-, 3-, 5- and 7-fold mole excess to bGH respectively. From Brems [88], reproduced with permission

corresponding to the residues 96–133 is incorporated in the folding mixture [88]. The helical peptide presumably interacts stereospecifically with the exposed hydrophobic parts of the corresponding helix of the monomeric intermediate. This interaction prevents association of two intermediates, and hence aggregation. Aggregation can be considerably suppressed by increasing the concentration of helical peptide (Fig. 6).

Two important conclusions can be drawn from the foregoing experimental findings of different researchers: (a) the in vitro refolding process yields relatively lower amounts of native protein, which are further decreased with increase in the initial concentration of denatured polypeptide, and (b) the decrease in yield is correlated with an increase in the inactive aggregates formed by the specific association of partially folded intermediates. The folding steps follow first order kinetics with respect to the concentration of denatured polypeptide. On the other hand, aggregation follows second or higher order kinetics. There is a kinetic competition between aggregation and folding. The rate of folding does not alter appreciably with an increase in the protein concentration, while aggregation rate does. Consequently, aggregation predominates over folding at higher protein concentration. The kinetic competition of a hypothetical folding and aggregation pathway is shown in Fig. 7. The extent of aggregation is believed to increase with increasing hydrophobicity of the protein, as well as with increase in the thiol groups. In order to avoid aggregation and to maximize yield, protein folding is conducted at lower concentration of protein. Table 5 shows the concentration of proteins used in different refolding experiments.

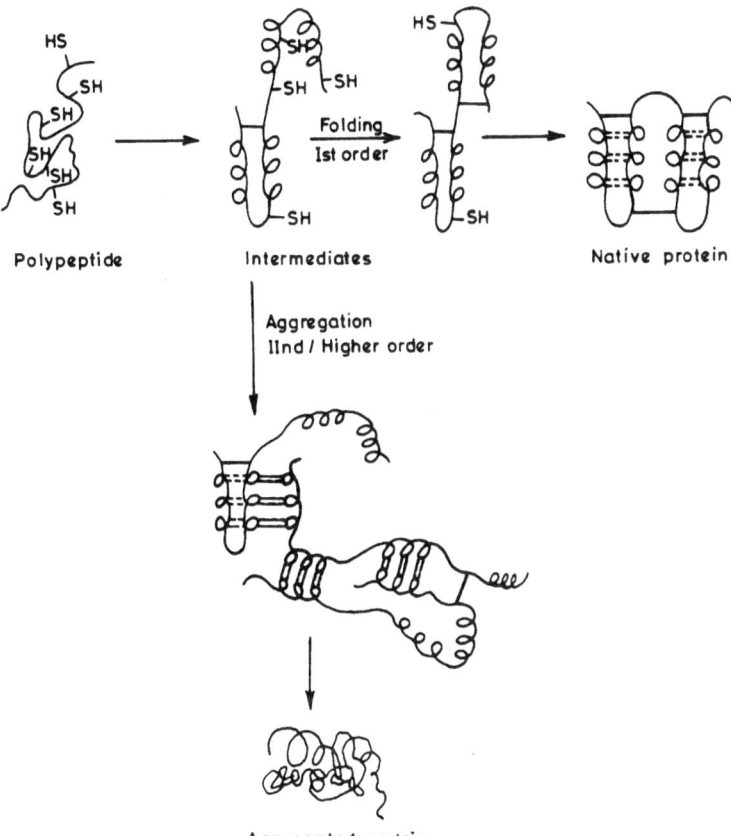

Fig. 7. The kinetic competition between hypothetical folding and aggregation pathways of a poly-peptide containing thiol groups. From Mitraki and King [14], reproduced with permission

5 Selection of Purification Protocol

Downstream processing of intracellular proteins encompasses a sequence of separation steps by which the active protein molecules are separated from proteinaceous and non-proteinaceous contaminants. There are four major steps involved in the purification of a protein, expressed in *E. coli* as inclusion bodies. These are:

1. isolation of inclusion bodies;
2. purification of protein in the denatured state;
3. refolding of protein into bioactive conformation; and
4. polishing of the bioactive molecules.

Table 5. Concentrations of proteins used in the different refolding experiments using heterologous proteins

Protein	Nature	Concentration ($\mu g\,ml^{-1}$)	Reference
1. M-CSF	Dimeric protein of 218 amino acids. 9 half-cystines, 4 inter-subunit disulfide bonds	700	[89]
2. t-PA	527 amino acids. 35 half-cystines, 17 disulfide bonds	100	[90]
3. Insulin	Chain-A (21 amino acids) and Chain-B (30 amino acids) 3 disulfide bonds	200	[91]
4. Bovine carbonic anhydrase B (CAB)	31 kDa	300	[92]
5. Human glia maturation factor beta (hGMFβ)	141 amino acids 1 disulfide bond	250	[93]
6. hG-CSF	175 amino acids 2 disulfide bonds	1000–2000	[94]
7. bGH	191 amino acids 2 disulfide bonds	180	[88]
8. IL-2	3 half-cystine, 1 disulfide bond	1	[95]
9. βhCG	145 amino acids 6 disulfide bonds	120	[83]
10. Fab-fragment	50 kDa	80	[82]

Before discussing the major purification steps in detail, it is important to answer the following questions in order to establish a rational approach of protein purification.

I. Whether upstream processes are standardized? Purification procedures cannot be optimized unless the fermentation processes are standardized. The amounts of host cell contaminants in the crude target protein primarily depend on the mode of its cultivation, medium composition, and the time of harvest.

II. What is known about the physico-chemical properties of the target protein? Knowledge on the physico-chemical properties of the target protein are most essential for the design and selection of powerful purification techniques. Important among them are molecular size, charge, isoelectric point, solubility, surface hydrophobicity, chelating property, antigenicity, pH and temperature stability, stability and dependency on cofactors and ions, stability in organic solvents, surface distribution of lipophilic nonpolar residues, distribution of amino acids on the surface of the folded protein, number of free thiol groups and disulfide bonds etc. Besides, knowledge of sensitivity of the target protein to the host proteolytic enzymes is also important.

III. What is known about the contaminants? The objective of purification is to separate large amounts of contaminating matters from the target protein. It is therefore important to know the nature and the concentration of those contaminants present.

IV. Whether proper assay systems are available? It is of paramount importance to know the initial quantity/activity of the target protein and how much is present at the end of each purification step. Without these the efficiency of each process step cannot be determined, and so optimization is not achieved. It is also necessary to have sensitive assay systems to determine traces of DNA and endotoxin present in the final product.

5.1 Isolation of Inclusion Bodies

This step is composed of two main processes, namely the separation of inclusion bodies from the host cells and solubilization of inclusion bodies. There are many ways by which living organisms are disrupted to release intracellular products. The disruption procedure suitable for one kind of cell may not be suitable for other kinds. Guidelines for the selection of disruption techniques amenable to various living organisms have been elaborated by Hopkins [96]. Details of these techniques are beyond the scope of this review. Here, discussion is limited to the techniques of disruption of *E. coli* cells, and separation of inclusion bodies.

The outer membrane of *E. coli* is composed of lipoproteins, lipopolysaccharides, and proteins, whereas phospholipids and proteins are two components of the inner membrane [97]. In addition, the periplasmic space contains peptidoglycan. These multilayer lipid-protein-glycan complexes contribute to the rigidity of the *E. coli* cell wall more than any other organism. Due to this rigid cell wall, inclusion bodies produced in *E. coli* cannot be isolated satisfactorily by techniques like ultrasonication, osmotic shock, chemical permeabilization, and enzymatic lysis. Though ultrasonic disintegration is the most widely used technique, the disruption of bacterial cells is reported much lower than 100%. For all practical reasons, there is a limitation to the scale-up of the batch-type ultrasonicator. Continuous flow ultrasonicators are available from several manufacturers (Braun, Bronson, Sonic Systems); however, the disruption of yeast cells has been reported to be only 60% in these systems [98]. Osmotic shock and enzymatic lysis processes are limited to the periplasmic and membrane bound proteins, as the cells and the inner membrane remain unaffected in these treatments [99]. Chemical permeabilization using a mixture of 0.5% Triton X-100 and 0.1 mol/l GuHCl releases about 53% of intracellular proteins [100]; however, the method seems to be inefficient for the extraction of inclusion bodies from the host cells.

The best technique for the disintegration of *E. coli* cells is high pressure homogenization, though it suffers from the drawback of contamination of target protein with the cellular components. The problem of cellular contaminants in the target protein is not severe in the case of inclusion bodies, as they are heavier than cellular fragments. The main advantage of using a high pressure homogenizer is that complete disintegration of the cell wall is possible in a short time and the process is easily scalable up to the commercial level. The most familiar laboratory scale high pressure homogenizer is the French Press (SLM Instruments, Urbana, IL), where pressures as high as 2500 kg cm^{-2} can be achieved on the cell suspension. The Manton Gaulin-APV homogenizer (APV Gaulin, Everett, MA) is the most documented large scale system. The system can be used for continuous operation, and its capacity varies from 55 to 4500 l h^{-1} at cell concentrations of 10–17% (w/v).

The working principle of the high pressure homogenizer is to force the cell suspension through a narrow channel or orifice under high pressure. The disruption of the cell wall occurs by a combination of the large pressure drop and strong shearing force. The design details of high pressure homogenizers are given elsewhere [101]. The performance of this system is a function of the operating pressure, valve design, number of passes through the valve, flow rate, and temperature of the cell suspension. The amounts of intracellular proteins present in the cell lysate determine the extent of disintegration. At a pressure of 600 kg cm^{-2} the amounts of intracellular proteins released are increased with the number of passes. Eighty two percent of the intracellular proteins are reported to be released within three passes in a Manton Gaulin homogenizer [102]. The extent of disruption is higher in induced cells than in uninduced ones. Middelberg et al. [103] have reported more disruption in the induced E. coli expressing inclusion body of porcine somatotropin than in the uninduced cells. For the same system the extent of disruption reduces with increase in cell concentration in the feed. The disruptability of cells is also influenced by the physiological state of the cell growth. Cells in the exponential phase of growth tend to produce thinner walls, which are easily disruptable [96].

The inclusion bodies are separated from the cellular fragments by centrifugation at 5000 to 10 000 (x g) for a period of 5 to 10 min. The duration and speed of centrifugation depends on the difference in densities between inclusion bodies and the cellular fragments. Continuous flow centrifuge (intermittent nozzle discharge type) is commonly used in industrial scale processing of inclusion bodies. The flow of the process stream, speed of centrifugation, physical properties of the sedimenting particles, and density and viscosity of the solution phase are important to determine the efficiency of centrifugation. The inclusion body particles settle as a pellet, which is further clarified by washing and centrifugation. The major associated cellular contaminants are removed by 3–4 cycles of washing, the finish of washing being determined by examining the pellet suspension under an inverted microscope. The inclusion body pellets of hGH and interferones have been purified up to 90 and 50% homogeneities, respectively using the multiple washing technique [79]. Washing with lower concentrations of denaturating agent without solubilizing the inclusion body helps to improve the purity of the target protein. The inclusion body of human proinsulin has been purified from 25% total stainable protein in the cell homogenate to 75% by washing with 3 mol/l GuHCl, followed by low speed centrifugation [91]. Often membrane associated proteins and lipids contaminate the inclusion body pellet. These are selectively extracted either by treatment with a neutral

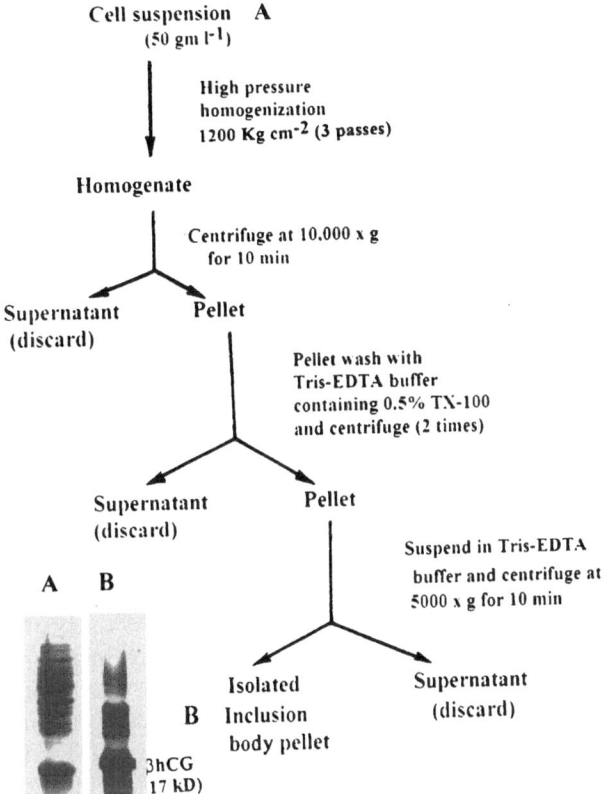

Fig. 8. Scheme for the isolation of βhCG inclusion bodies expressed in *E. coli*. A: 12.5% (w/v) SDS-PAGE of total cell lysate (12% βhCG on the basis of total stainable proteins); B: 12.5% SDS-PAGE of isolated inclusion body pellet (50% βhCG on the basis of total stainable proteins)

detergent like Triton X-100, or sodium deoxycholate [104]. This treatment not only removes contaminants but also minimizes the proteolysis of the target protein by the membrane associated enzymes.

Figure 8 shows the detail of a scheme for the isolation of βhCG inclusion bodies expressed in *E. coli* BL21 by plasmid pVEX11 under the control of phase T7 promoter. Wet cell pellet is suspended in 50 mmol/l tris-HCl buffer (pH 8.0) containing 5 mmol/l EDTA. EDTA destabilizes the outer membranes of *E. coli* and renders them more fragile to the shear forces [105]. Inclusion body of βhCG has been purified from 12 to 50% total stainable protein using this isolation scheme.

Dense structure of inclusion body is solubilized by minimizing intermolecular hydrophobic interactions and the formation of aberrant disulfide bonds. Solvents used for solubilization are selected on the basis of the criteria that it should not create difficulties in the final step of purification, readily solubilize inclusion bodies, inhibit proteolytic activity, and not chemically modify labile amino acids in the protein.

Detergents of any form (cationic, anionic and zwitterionic) have been used for the solubilization of inclusion bodies. At low detergent concentration, the hydrocarbon chain of the detergent interacts with the side chain of hydrophobic amino acids, and the detergent polar group extends out into the bulk water. Thus, detergent just masks certain hydrophobic patches on the protein surface and enhances its solubility. Detergents are relatively cheap, and the solubilized proteins often display full biological activity after refolding. The major drawback of using detergents is its complete removal from the target protein. In certain cases they are retained by the chromatography matrices and ultrafiltration membrane. Detergents also solubilize membrane proteases along with the inclusion bodies, which lead to the poor yield of the target protein. Sodium dodecyl sulfate (SDS) has been used for the solubilization of interleukin 2 (IL-2), IFNβ [106] and bGH [107]. Mild detergent, namely n-lauryl sarcosine, is also reported to be useful for the solubilization of inclusion bodies [108].

Chaotropic agents, such as GuHCl and urea, are commonly used for the solubilization of inclusion bodies. The extent of solubilization depends on the concentrations of protein and denaturant, purity of the protein, pH, and presence of free thiol groups. The concentration of chaotropic agent required to solubilize inclusion bodies increases with increase in the hydrophobicity of the protein. GuHCl and urea at a concentrations of 6 and 8 mol/l respectively are sufficient to solubilize inclusion bodies. Often it is difficult to solubilize inclusion bodies even at higher concentration of denaturant. This is mostly due to the formation of aberrant disulfide bonds (intermolecular). In these cases proteins are solubilized by reducing disulfide bonds with β-mercaptoethanol or DTT. It has been observed in the author's laboratory that the inclusion bodies of βhCG turned into a mass with rubber-like texture during isolation. This material could be dissolved in 6 mol/l GuHCl containing reducing agent. Urea, although less effective, is used at a higher concentration in place of GuHCl, and is much cheaper. The drawback with urea is that it spontaneously forms cyanate ions at higher concentration and at high pH, particularly when the temperature is more than 10–15°C. This cyanate ion modifies protein amino groups and increases its heterogeneity. The problem of cyanate formation is minimized by using urea in presence of tris-HCl buffer.

Extreme pH buffers are cheapest and effective solvents for inclusion bodies, but these irreversibly damage the proteins. Acetic and butyric acid (5–80%) have been used to solubilize IL-2 and IFNβ respectively [109, 110]. Inclusion body of bGH has been solubilized at a pH higher than 11 [111].

The extent of solubilization of inclusion bodies in GuHCl, urea, and in alkaline pH buffer can be monitored by measuring intensities of fluorescence spectrum and perpendicular scattered light signal of the solution. Both the intensities are found to fall with an increase in the denaturant concentration, and hence with the increase in the solubility of the inclusion bodies, as found in the case of prochymosin [112]. The emission maxima of fluorescence spectrum in these systems are also shifted to the region of higher wave lengths. The concentration of denaturant at which the intensity of spectrum is recorded at minimum represents the point of maximum solubility [112]. The amount of denaturant required to solubilize inclusion bodies can be optimized by these techniques. Fluorescence spectra are also used by the other investigators to determine the solubility of the inclusion bodies [113].

5.2 Purification of Protein in the Denatured State

Inclusion body pellets consist of varying proportions of impurities (e.g., proteins, DNA and lipids) of host origin. The purification technique to be followed prior to refolding of the target protein is decided by the consequences of these impurities on the refolding process. It has been reported that the addition of crude E. coli proteins or pure albumin to the refolding mixture does not alter the renaturation yield of protein [84]. Purification of target protein prior to the folding is governed by the following two factors.

(A) Expression level and the purity of the target protein in the inclusion body pellet – purification of a protein prior to refolding is recommended, if the expression level is low (< 15% of the total cellular proteins), and/or the purity of the target protein in the inclusion body pellet does not achieve more than 40–50% of the total stainable proteins. However, prior purification has been shown to incur additional expenditure without much benefit to the refolding process, where the target protein in the inclusion body pellet is in a highly purified form (80–90%).
(B) Thiol containing contaminating proteins – the presence of reactive protein contaminant (containing -SH group) may lead to the formation of inter-molecular disulfide bonds with the target protein. As a result, the yield of the bioactive molecule is reduced. Hence, the purification of target protein from the thiol containing protein contaminants prior to refolding is beneficial.

The purification of protein in the denatured state offers access to some of the powerful chromatography techniques; these are as follows.
Reversed phase chromatography-RP-HPLC is the most powerful chromatography technique which could be used in the case of denatured protein. The peak resolution is excellent and about 98% purification can be achieved in one step. In the case of a hydrophobic protein, the yield can be improved further by incorporating suitable denaturant in the mobile phase [114].
Gel filtration – size exclusion chromatography is often used to purify protein in the denatured state. The aggregation of protein and protein-matrix interaction are prevented in denatured protein solution. Thus, better separation of the target protein and its higher yield is achieved [115]. Gel filtration matrices of Sephacryl series (Pharmacia Fine Chemicals, Sweden) and Bio-Gel P series (Biorad, USA) are most stable in 6 mol/l GuHCl and in 8 mol/l urea, and so can be used repeatedly without compromising the resolution power.
Ion exchange chromatography – the method is useful for the purification of denatured protein from a dilute solution. Both cationic and anionic matrices can

be deployed on the basis of isoelectric pH and the pH stability of the target protein. Most popularly used stable matrices are Q- and S-Sepharose (Pharmacia Fine Chemicals, Sweden), and Macro-Prep 50Q and 50S (Biorad, USA). Ion exchange chromatography offers easy scalability, and both concentration and purification of the target protein are achieved in a single step. Protein denatured in GuHCl is not applied to the ion exchange column: instead, GuHCl is exchanged with a suitable concentration of urea. Viral coat proteins of Sendai virus [116] and subunits of calf brain tubulin [117] have been purified in the denatured state using ion exchange chromatography. Recently βhCG has been purified from 30 to 85% homogeneity by single step Q-Sepharose ion-exchange chromatography in 3 mol/l urea-tris.HCl buffer [83].

The target protein can be clarified to a considerable extent by high speed centrifugation or by cross flow filtration (ultrafiltration). The later is applicable to large scale operations. Most of the recombinant proteins produced in *E. coli* are within the molecular mass range of 14–22 kDa. By careful selection of the membrane molecular weight cut-off (30 000 MWCO in these cases), it is possible to separate the majority of the macromolecular contaminants from the target protein.

5.3 Refolding of Proteins into Bioactive Conformation

To obtain native conformation, the polypeptide chain has to be refolded into correct secondary and tertiary structures, which are further stabilized by the formation of intramolecular disulfide bonds. In order to facilitate the same, in vitro refolding strategies are developed on the basis of the following conditions: (a) the concentration of denaturant is reduced to a level at which intramolecular stabilizing forces (e.g., hydrogen bonding, hydrophobic interactions) do exist, (b) the folding process is carried out in a buffer containing oxidizing agent, and (c) concentration of denatured protein in the refolding buffer is maintained low to avoid intermolecular aggregation.

5.3.1 Refolding by Dialysis, Diafiltration and Gel Filtration

These are the most frequently employed techniques for the exchange of buffer. In dialysis and diafiltration, membranes of defined molecular weight cut-off are used. Since the membrane molecular weight cut-off is much lower than the target protein, it is retained by the membrane, whereas buffer exchange results in refolding of protein. In gel filtration, the denaturant enters the pores of the matrices, whereas protein is exposed to the refolding buffer. This method of buffer-exchange is much faster than dialysis, and the relative solubilities of the folding intermediates determine the success of the technique [118]. The refolding of proteins by these methods is done only when the intermediates of the protein folding pathway are soluble in the native environment. A number of patents have been claimed for the refolding of prochymocine, bGH, IFNγ by dialysis and diafiltration [118].

5.3.2 Refolding by Dilution

5.3.2.1 Refolding by Single Stage Dilution

This method is less time consuming than the previous ones. The denatured inclusion body is diluted 25 to 50 times in the refolding buffer, whereby the concentration of denaturant is reduced and the native secondary structures of protein are restored by various stabilizing forces. At the same time the formation of aggregate is minimized in dilute solution of protein. The dilution is carried out by (a) slowly adding denatured protein to the refolding buffer, or (b) quickly pouring refolding buffer into the denatured protein solution.

The refolding buffer is typically composed of 50 mmol/l tris-HCl (pH 8–9) containing different redox systems, such as reduced/oxidized glutathione or atmospheric oxygen (air oxidation). In air oxidation the thiol residues of the protein are oxidized by soluble oxygen. Oxidation is performed by aeration of the protein solution. Air oxidation is catalyzed by incorporating 0.1–1.0 μmol/l Cu^{+2} ions in the refolding buffer. The formation of disulfide bonds in air oxidation has been found more effective in the presence of trace amounts of thiol agents, e.g., 2-mercaptoethanol, DTT, or cysteine. Air oxidation technique has been used for the refolding of IL-2 [119], bGH [120], and pro-urokinase [121]. Air mediated oxidative refolding is the cheapest alternative, but a slower process. In general the refolding of protein is carried out for 24 h. The time required for the formation of 50% of two disulfide bonds in G-CSF ranges from 0.5 to 4.5 h [122]. This method of disulfide bond formation is not suitable in cases of proteins which, in the reduced state prior to oxidation, do not attain stable near-native conformation [122]. Also, oxygen mediated disulfide bond formation is found unsuitable in cases where native proteins contain one or more free cysteine residues [123]. This method is only suitable for proteins containing one or two disulfide bonds, and possessing relatively simple refolding requirements.

The oxidation of thiol groups to disulfide can be controlled in the presence of a suitable redox buffer (reduced/oxidized glutathione). The rate of oxidation in redox buffer system is much faster than air oxidation. The kinetics of the thiol-disulfide exchange process in controlled by the concentrations of reduced and oxidized glutathione and their molar ratios to the protein-cysteine concentration. As a result of controlled oxidation, the yield of the refolded protein in this case is found to be much higher than in the air oxidation process. Reduced and the oxidized glutathione at a molar ratio of 10:1 is generally used. A number of proteins, such as t-PA [124] and M-CSF [125], are reported to form correct disulfide pairings in reduced/oxidized glutathione buffer. The processes for the refolding of IFNγ, IL-2, t-PA, and M-CSF by single stage dilution in redox buffer have been patented by various researchers [118]. The main drawback in this refolding technique is that the process ends with dilute protein solution of large volume.

5.3.2.2 Refolding by Stage Dilution

The refolding of proteins by stage dilution is most popular amongst the alternative strategies documented in the literature. A series of patents has been filed for various recombinant proteins refolded by this technique [79, 126, 127]. Initially, inclusion body is dissolved in strong denaturant (6 mol/l GuHCl or in 8 mol/l urea), and then the concentration of denaturant is reduced stage-wise whilst maintaining the protein in solution. Thiol-disulfide exchange is carried out in the intermediate denaturant concentration. At the end of refolding the denaturant is removed by dialysis or diafiltration.

This approach to refolding is most suitable where reduced form and/or folding intermediates of the protein are insoluble in the native environment. Denatured and reduced chymotrypsinogen has been successfully refolded by diluting in two stages [128]. In the first stage the concentration of GuHCl is brought down from 6 mol/l to 1 mol/l, where thiol-disulfide exchange is carried out. The intermediates of the folding pathway of chymotrypsinogen are believed to be stable in 1 mol/l GuHCl. Later, the remaining GuHCl is dialysed.

It is necessary to know the concentration of denaturant at which folding should be carried out. This is to avoid the conformational restriction (at high denaturant concentration) and inaccessibility of the thiolate ions to the disulfide reagent (at low denaturant concentration), and to maintain the soluble form of the intermediates. The refolding conditions can be approximated in a transverse urea gradient gel electrophoresis (Fig. 9). Electrophoresis is carried out with the

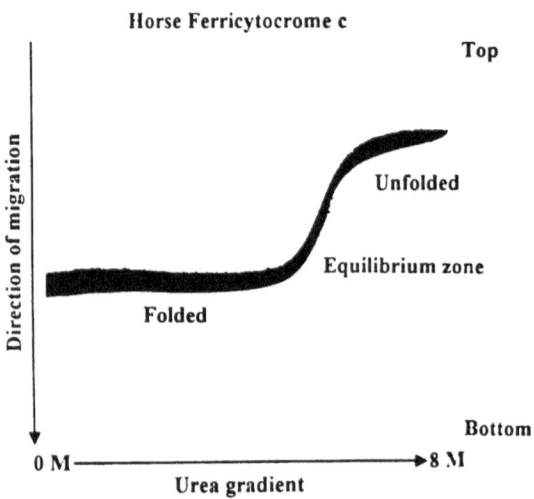

Fig. 9. Transverse urea gradient (0–8 mol/l) electrophoresis of horse ferricytocrome c. The *stained protein band* shows the presence of three distinct zones corresponding to unfolded, equilibrium and folded state of the protein. From Creighton [129], reproduced with permission

denatured protein in 8 mol/l urea. After electrophoresis, the gel is fixed and stained for protein. Three distinct zones of the protein band may be seen in the gel [129]. Unfolded protein molecules have a larger Stokes radius than the folded molecules, so migration is retarded. On the other hand, the folded protein is much more advanced in the gel. In horse ferricytochrome c, the unfolded and folded states of the protein molecules are in equilibrium in 4–6 mol/l urea (Fig. 9). Before in vitro oxidative folding is initiated, the concentration of urea in the protein solution is diluted near to the upper limit of folding, in this case 4–5 mol/l.

Better refolding yield may be obtained in cases where thiol groups are reversibly protected by sulfitolysis. Denatured protein (inclusion body) is treated with sodium sulfite and sodium tetrathionate [79]. Sodium tetrathionate generates disulfide bonds from any thiol groups, and the sulfite ions break the disulfide bonds by nucleophilic attack. Sulfite ion substitutes one of the sulfide partners, and thus the sulfonated protein ($P\text{-}SSO_3$) is formed. In the refolding process, the sulfonate group is nucleophilically displaced by ionized sulfhydryl compound (RS^-) and forms the mixed disulfide. Mixed disulfide is converted into disulfide by intramolecular displacement. The chemical reactions for the formation of protein disulfide from a sulfonated protein are shown in a recent paper [83]. The process of reversible protection of thiol groups prior to oxidative refolding has been found effective in cases of insulin [91], t-PA [130], and βhCG [83].

5.3.3 Additives-Assisted Refolding

In vitro refolding efficiencies of proteins have been improved by incorporating various additives in the refolding buffer [7]. It is believed that the additives either bind to the folding intermediates and inhibit their aggregation, or stabilize the refolded native structure of the protein, and thus the yield is substantially improved. Amino acids, sugars, neutral surfactants, and polymers belong to the category of additives used, which can be easily separated from the refolded protein. Initial work on additive (0.2 mol/l arginine) mediated refolding of recombinant prourokinase has been carried out by Winkler and Blaber [131]. Later, Jaenicke and Rudolph [132] have shown that L-arginine acts as a labilizing agent, which preferentially destabilizes the wrongly folded proteins. The correctly folded protein molecules remain unaffected, whereas the molecules trapped in the non-productive side reactions are reshuffled. Addition of 0.4 mol/l L-arginine in the refolding buffer has been found to improve refolding efficiency of recombinant Fab-fragments by 60% [82]. Similarly, 0.1 mol/l glycine improves the refolding yield of relaxin [133]. Sucrose and glycerol at higher concentrations (> 10% w/v) in the refolding mixture have been shown to improve the refolding efficiency of β-lactamase [134] and lysozyme [135] respectively. Polyethylene glycol (3500 MW) has been successfully utilized for improving yields of recombinant RNaseA, IFN_γ and t-PA [136].

5.3.4 Chemical modification

The free amino group (lysine, arginine and N-terminal) of a protein is reversibly modified to anionic carboxylic group by treatment with citraconyl anhydride. The resulting polyanion protein molecules repel each other and thus prevent aggregation in aqueous buffer. In the case of disulfide containing proteins, the thiol groups are sulfonated prior to the acylation. Once a disulfide bond is formed, the protein is deacylated and the native structure is regained by lowering pH of the aqueous buffer to 5. The protein refolded by this technique has been found to attain biologically active conformation. Recombinant HIV envelope protein has been refolded to bioactive molecules by reversible acylation [137]. Despite the success in processing inclusion body through N-acylation, the technique has not become popular in the area of therapeutic proteins, probably due to incomplete deacylation. For applications of in vitro diagnostics, the protein refolded through chemical modification could be useful.

5.3.5 Refolding in Reversed Micelles

In this case, each denatured protein molecule is encapsulated in an aqueous droplet stabilized by the surfactant, and the micelles are dispersed in an organic solvent. As the micelles do not coalesce, the protein is not aggregated upon refolding. The structure of the reversed micelles are shown in Fig. 10A. By selective exchange process the concentration of denaturant in the micelles is reduced, and the protein is allowed to refold in presence of redox buffer. Protein is refolded inside micelles whereby it attains biologically active conformation (Fig. 10B). After refolding the protein is extracted into an aqueous buffer (Fig. 10C). Hagen et al. [138] demonstrated successful refolding of RNaseA in reversed micelles using an AOT (di-2-ethylhexyl sulfosuccinate) and isooctane system.

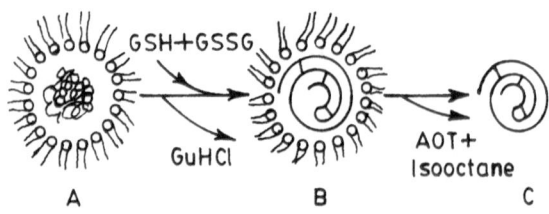

Fig. 10A–C. Folding of protein in the reversed micelles: **A** Structure of reversed micelles where denatured protein molecule is surrounded by surfactant (AOT) and the polar groups of the surfactants are towards the protein and nonpolar chains are on the bulk of the fluid (isooctane); **B** refolded protein molecule in the reversed micelles at low concentration of denaturant; **C** extraction of refolded protein in the aqueous buffer (breaking of reversed micelles)

5.4 Polishing of the Bioactive Molecules

Purification of protein to the level of therapeutic grade requires a battery of sophisticated chromatography steps. Selection of these steps is dependent on (a) concentration and purity of the target protein, (b) its physico-chemical properties and stability, (c) required purity level in the final product, and (d) load of contaminants, and their permissible limits in the final product. The purity of the target protein subjected to the polishing steps is generally greater than 90%. The remainder is both proteinaceous and non-proteinaceous contaminants. The objective at this stage of purification is to remove these low concentrations of contaminants from the target protein.

The range of chromatography techniques that could be used at these stages of purification are listed in Table 6. Since some techniques are better suited to the earlier stage of purification, while others are better suited to the final stage, it is important to consider carefully the proper sequence of chromatography steps. A judicious sequence reduces the number of buffer exchange steps, improves yield, speeds-up the overall purification process, and reduces costs. It is revealed

Table 6. Application of different chromatography techniques in the polishing of bioactive molecules

Type	Resolution	Speed	Capacity	Cost	Applicability
Gel Filtration (GF)	C	C	C	B	Suitable for final stage, remove oligomeric proteins and ligands used in the affinity based purification
Ion Exchange (IEC)	B	A	A	B	Suitable for early and intermediate stages, sample should have low ionic strength and correct pH
Hydrophobic Interaction (HIC)	B	A	A	B	Suitable for early and intermediate stages, ionic strength of sample is maintained high (may be used after IEC and AFC)
Reversed Phase (RP-HPLC)	A	B	B	A	Final stage for purifying low molecular weight protein contaminants, disulfide isomers, deaminated proteins etc.
Affinity (AFC)	A	A	A	A	Suitable for early and intermediate stages, sample should be at low ionic strength, proper ligand binding conditions required

Resolution: D-Poor; C-Moderate; B-High; A-Excellent
Speed: C-Low; B-Moderate; A-Fast
Capacity: C-Low; B-Moderate; A-High
Cost: B-Moderate; A-High

from an analysis that in most of the cases ion exchange chromatography has been used as the first step of the final purification, which is followed by affinity and gel filtration chromatography steps [139]. However, there are exceptions, where a different sequence of steps yielded better results. Human leukocyte interferon has been purified over 1000-fold using an immunoaffinity step, followed by copper chelate affinity chromatography, ion exchange, and finally gel filtration [140, 141].

Anion exchange chromatography is most popular in the purification of inclusion body derived proteins. Beside its high resolution power and large throughput, it removes deleterious impurities like DNA and endotoxin. These impurities possess strong negative charges in the alkaline pH range and are thereby removed in anion exchange chromatography. Inclusion of specific steps to remove DNA and endotoxin in the downstream processing scheme are proved unnecessary in many cases. Anion exchange chromatography has been recommended for repeated use in protein binding and non-binding conditions [142]. Table 7 shows the performance of different anion exchange matrices in the removal of DNA and endotoxin from the protein solution. It may be noted that strong anion exchanger is most effective in the clearance of both DNA and endotoxin from the sample stream.

Various disulfide isomers are formed during refolding of thiol containing proteins (Table 2). Again, proteins undergo different modifications during in vivo and in vitro processing steps, and thus generate unwanted forms (e.g., deaminated and proteolytically degraded, etc.) of the target protein. These forms of the protein are separated in an RP-HPLC. Despite having tremendous resolution power, RP-HPLC is not yet popular in process scale purification of proteins due to partial or total loss of bioactivity. The bioactivities of IL-2 [143] and IFNβ [144] are reported to be reduced during purification in RP-HPLC. In process development HIC is often preferred over RP-HPLC because purified proteins retain better bioactivity due to its less harsh operating conditions. The operation in HIC is also simple. The successful report so far available in the application of RP-HPLC is the process scale purification of recombinant insulin [145].

Table 7. Performance of different anion exchange matrices for the removal of DNA and endotoxin from protein solution

Typical functional group	Log reduction	
	DNA	Endotoxin
PEI (weak)	1.0	2.0
DEAE (moderate)	1.0	0.5
QAE (strong)	3.0	2.0

Protein non-binding conditions followed. Protein solutions have been spiked with 10 μg of genomic DNA (mammalian cell) per mg protein, and 800 EU mg^{-1} protein [142]

6 Process Development and Scale Up

In the research stage of protein purification, product purity and characterization are of primary concern, while manufacturing costs, and process scalability are of secondary importance. The process development work is initiated once the product is passed through the initial stage of research. Process development in small scale production is mainly concerned with the selection of the individual process steps, optimization of operating conditions in each step of the purification, and process validation in terms of the efficiency of purification and purity, and the safety of the product. The chromatography matrices are selected at this stage on the basis of their selectivity, resolving power, capacity, and recovery efficiency. Preliminary costing of the purification process is also done at this stage. Another objective at this stage is to produce sufficient quantities of the product to carry out clinical trials. Full scale manufacturing of product is considered in the second stage of process development. It involves scale up, implementation of process, and regulatory aspects.

A review of large scale chromatography purification reveals that two basic approaches are followed in scaling up [145]. The first approach is primarily based on the large scale version of the analytical separation of protein. In this case scale up is done simply by increasing the column and sample volume in direct proportion to each other. The demerits of this approach are under-utilization of capacity and escalation of cost. In the second approach the operating conditions optimized in the process development steps during small scale production are utilized. The second approach of scale up renders maximum productivity and optimal use of the production set up.

7 Removal of Extra Methionine from N-Terminus of Protein

Recombinant proteins expressed in *E. coli* may contain an extra methionine at their N-terminal as a result of translation-initiation by *N*-formylmethionine residue. Though *E. coli* has the enzymatic capabilities in the cyotosol to deformylate (by deformylase) and subsequently remove this additional methionine (by methionine aminopeptidase), the removal efficiency is low in the heterologous proteins [146]. This extra methionine may affect the biological activity, and at the same time it may potentially increase the immunogenicity of the molecule. These abnormal functions are of particular concern in the clinical application of the proteins [147]. Therefore, it is necessary to remove this extra methionine from the heterologous proteins.

In vivo N-terminal processing of methionine by methionine amino-peptidase is substantially influenced by the nature of the adjacent amino acid residue [148, 149]. Methionine is not cleaved if the side chain of the amino acid next to it is large, or the amino acid is charged [146]. Incomplete processing of extra methionine is found when the second amino acid is of intermediate size, and/or the heterologous protein forms highly condensed structures. A partial removal of methionine in IL-2 has been reported, despite the presence of favorable amino acids adjacent to the methionine (-Ala-Pro-) for the action of methionine aminopeptidase [146]. It is presumably due to inaccessibility of the enzyme to the reaction site of the aggregated protein. On the other hand, in βhCG hormone subunit a complete removal of methionine has been reported [83]. This could be due to the presence of favorable amino acids next to the methionine (Ser-Lys-) and less aggregated form of the protein.

How this methionine is removed in vitro, has been elaborated by Ben-Bassat [146]. In this review a specific example for the removal of methionine (Novo Nordisk's Procedure) has been described. Human growth hormone is produced as a fusion protein with small N-terminal extension (Met-Glu-Ala-Glu-hGH). The fusion protein is purified and the extra peptide is specifically cleaved with dipeptidyl aminopeptidase I (DAP I, EC 3.4.14.1). The enzymatic cleavage is specific, DAP I cannot remove the Pro-Thr peptidyl bond [150]. Thus stepwise removal of two dipeptides (Met-Glu and Ala-Glu) produces authentic hGH (Phe-Pro-Thr-) of 22 kDa molecular mass.

8 Characterization of Recombinant Products

Rigorous characterization is a prerequisite for a purified protein to obtain permission from the regulatory authorities, prior to the induction in the phase of clinical trials. The proteins are characterized with the help of chemical, instrumental, and biological methods. A major advantage in rDNA technology is that protein expresses under the control of the genetic code, therefore its primary structure is assumed unchanged. Any alteration in the DNA level (plasmid DNA) may lead to modification of the protein. To avoid this change the original transformed culture (characterized at molecular as well as product level), known as master culture, is stored at − 70 °C under controlled conditions. The working seed lots and subsequent production cultures are sequentially prepared from the master culture. One may expect no change in the culture within limited generation numbers during preparation of the culture stock. Therefore, in routine control of the bulk final processed product, consistency, identity, purity, potency, and toxicity are checked using limited instrumental and biological methods. On the other hand, a new product is characterized in detail. Table 8 shows the general tests schedule for the characterization of a recombinant protein, intended to be used for vaccine and non-vaccine applications. The pure product is analyzed in light of its structure (primary, secondary, tertiary, and quaternary), function (immunological and biological), pharmacology, and toxicity.

Table 8. Tests schedule for the characterization of recombinant products

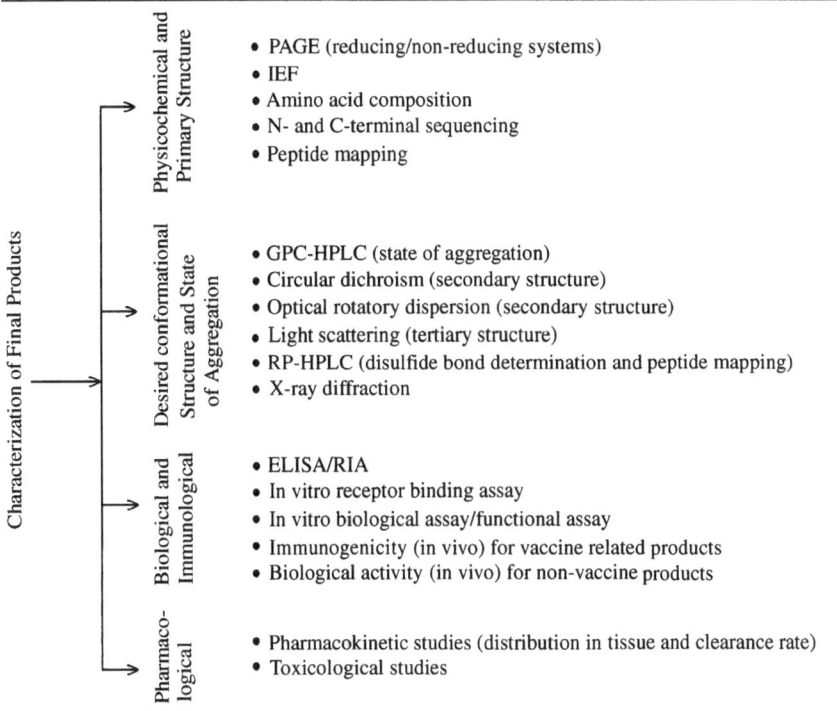

Characterization of Final Products		
Physicochemical and Primary Structure	• PAGE (reducing/non-reducing systems) • IEF • Amino acid composition • N- and C-terminal sequencing • Peptide mapping	
Desired conformational Structure and State of Aggregation	• GPC-HPLC (state of aggregation) • Circular dichroism (secondary structure) • Optical rotatory dispersion (secondary structure) • Light scattering (tertiary structure) • RP-HPLC (disulfide bond determination and peptide mapping) • X-ray diffraction	
Biological and Immunological	• ELISA/RIA • In vitro receptor binding assay • In vitro biological assay/functional assay • Immunogenicity (in vivo) for vaccine related products • Biological activity (in vivo) for non-vaccine products	
Pharmaco-logical	• Pharmacokinetic studies (distribution in tissue and clearance rate) • Toxicological studies	

The use of appropriate reference standards plays a crucial role in the characterization of a recombinant product. Except in a few cases, reference standards are prepared from natural sources. Many such characterized standards are available from the World Health Organization (WHO) and National Institute of Health (NIH). The final product is characterized and compared with the reference standard against each test.

The physical characterization of recombinant proteins have been discussed in detail in the literature [151]. Here, only biological and immunological aspects have been elaborated. In the case of a vaccine product, the two most important functional properties are antigenicity and immunogenicity. Antigenicity determines the capacity of a protein (antigen) to recognize antibody generated against it, whereas an antigen is considered immunogenic if it induces the formation of specific antibody in vivo. The key test to determine antigenicity is a quantitative immunoassay using a panel of antibodies (if available), directed to different epitopes. This test assures the presence of high levels of epitopes on inclusion body derived antigen. In the study on immunogenicity, animals (rodent species) are immunized by specific dosage of the antigenic protein, and, in due course of time, the level of neutralizing antibodies available in the blood serum is determined. The neutralizing efficiency of an antibody is determined by its ability to prevent an antigen binding to its receptor, an essential prerequisite to carrying out certain biological functions. This test can be carried out in vitro. For example, hCG binds to the LH/hCG receptor on the corpus luteum and supports production of ovarian progesterone essential to maintain pregnancy during the first two months. Antibodies generated against βhCG neutralize hCG and thereby prevent implantation of the fertilized ovum in the uterus, and the pregnancy does not occur [152]. The neutralizing capacity of βhCG antibody in vitro is determined by the direct binding assay of hCG to LH/hCG receptor on Leydig cells. The βhCG antibody forms immune complexes with hCG molecules and thus free hCG is not available to the receptor. In vivo immunogenicity of an antigenic protein against certain infectious diseases is performed by immunizing non-human primates (chimpanzees, monkeys) for a specified period of time, followed by challenge with the virulent infectious organism. Immune protection is ascertained by the survival of the animals following such a challenge.

In non-vaccine proteins, in vitro and in vivo bioassays are most important. In vitro assay determines the potency of the product. It does not require any living animal, and is based on the chemical action of a biological product. This assay is much faster than cell culture based bioassays and animal model assays (in vivo). The activity of t-PA is determined by in vitro clot lysis assay. A synthetic fibrin clot is formed in the presence of plasminogen from the action of the enzyme thrombin on fibrinogen. In presence of t-PA the plasminogen is converted to the active enzyme plasmin, which then induces lysis of the synthetic clot. The extent of lysis of clot is measured spectro-photometrically [151]. Similarly, in vitro assay of human RNaseA is performed by reacting it with yeast t-RNA under standard experimental conditions [153]. The increase in optical density at 260 nm is the measure of the RNaseA activity.

Cell culture based bioassay is less complicated than in vivo tests in animal model. This bioassay provides information on the effect of the product on the living system, hence results are considered more useful than those obtained in a simple in vitro assay. Antiviral activity of human IFNα can be measured in a simple cell culture experiment. The susceptible human lung carcinoma cell line is grown in a microtitre plate, prior to the incubation with different dilutions of IFNα. The cells are then challenged with encephalomyocarditis virus. The dilution of IFNα at which 50% protection of monolayer achieved is the measure of the antiviral activity of IFNα [151]. The biological activity of IL-3 is determined by the proliferation assay of IL-3 dependent cell (AML 193). The amount of [^{3}H]-thymidine incorporated by the cells is an indirect measure of the bioactivity of IL-3 [154]. Stimulation of testosterone in Leydig cells induced by hCG hormone is another example of cell culture based bioassay [155].

In vitro radio-receptor binding is another class of assay to determine the authenticity of the biological products. Each biological has a specific receptor, where it binds and transduces signals for biological functions. Natively folded protein only binds to its receptors; thus the assay is also a measure of the refolding efficiency. The assay is carried out as follows: the receptors are isolated from the specific cells and incubated with the serial dilutions of the test protein, say insulin, along with the standard radio-labeled [^{125}I]-insulin. The percentage binding of the labeled insulin will gradually decrease with the increase in the concentration of the test insulin in the solution. The binding of the labeled insulin to the receptors is measured in a γ-counter.

Finally, the bioefficacy of the product is determined in vivo. This is mandatory before any product is released in the market for clinical use. USP rabbit hypoglycemic and rat weight gain are two such examples of in vivo tests carried out for insulin and hGH respectively.

9 Purity and Safety

The purity of a biological may be defined as the measure of the active molecules in relation to the total substances (without additives) present in the product, and is generally expressed as a percentage. In biologicals the term potency is often used. It means biologically active form present in unit weight of the purified product. The purity of many rDNA-derived proteins may exceed 97%, with minimal impurities of host origin. This level of purity, which is achievable using sophisticated chromatography techniques, has been considered mandatory for rDNA products. There are no general guidelines of regulatory agencies on the acceptable purity level of rDNA products, and it varies from case to case [156]. Generally, a more conservative approach is adopted to evaluate the purity levels of health-care products. Regulatory agencies are more concerned with the impurities in the final product, whatever be the nature of it.

The nature and the type of impurities encountered in the preparation of a therapeutic grade protein from inclusion bodies are listed in Table 9. The impurities may originate from the host cells, or raw materials/chemicals used in the synthesis and purification steps. Impurities may also be added from the process water, equipment, and from the microbial/viral contaminations. The impurities in biological products have been classified into four types, namely innocuous, deleterious, major, and minor [157]. The innocuous impurities do not have much toxic effect, unless the drug is administered by a route other than that recommended. On the other hand, deleterious impurities may pose a major health hazard in terms of toxicity, carcinogenicity, and immunogenicity [151]. The regulatory agencies are more concerned with these deleterious impurities, and so their levels should be strictly controlled in the final product. The levels of major impurities may vary between 0.5–1%. It is recommended to gather detailed information on all the major impurities, pertaining to their toxicological, paratoxicological, and immunological properties. In case of minor impurities (< 0.5%), proper identification is only required.

The tests of pre-clinical safety and toxicity are mandatory for rDNA products to be used in humans for both vaccine and non-vaccine purposes. In cases where the test results of toxicity in animals are adverse, an attempt should be made to

Table 9. Nature of impurities and its consequences

Impurities	Endogenous (−)/ Exogenous (+) source	Medical/general consequences
1. Proteinaceous substances		
• host cell proteins	−	Unwanted immunological
• medium ingredients	+	responses and biological
• enzymes e.g. lysozyme, nucleases	+	activities
• antibody	+	
2. Target protein (abnormal forms)		
• aggregated/dimers etc.	−	Altered biological and
• deaminated	−	immunological activities
• disulfide isomers	−	
• proteolytically degraded forms	−	
• mutant forms	−	
3. Nucleic acids		Oncogenic response.
• DNA	−	Integration of foreign DNA
• RNA	−	into the genome of
		receipient cells may alter
		expression of cellular genes,
		and also express foreign gene
		products
4. Pyrogenic substances	− / +	Fever and septic shock,
		may lead to death
5. Chemical contaminants		
• antibiotics	+	Allergic actions
• buffer components	+	
• chaotropic agent	+	Irritant effect
• antifoaming agent	+	
• detergents	+	
• inducer	+	
• proteolytic inhibitors	+	Metabolic inhibition
• ligands and other chromatographic breakdown products	+	
6. Microorganisms	+	Potential for microbial infection

understand and to identify the reasons for such toxicity. If the toxic components are identified, they should be reduced further in purification steps to a level much lower than the toxic level. If necessary, the dose level of the product and its route of administration may need to be changed to minimize the toxic effect. The toxicity studies of recombinant insulin, interferons, and interleukin-2 have been elaborated in the literature [158]. Selection of appropriate animal species and the duration of the study are the two most significant factors for the assessment of the data relating to toxicity of the product.

The impurities are detected and quantified with the help of different analytical techniques as listed in Table 10. Intracellular products obtained from inclusion bodies are often associated with traces of DNA and pyrogenic substances, which belong to the class of deleterious impurities. The detection techniques used for

Table 10. Impurities, detection techniques and the permissible limits in the therapeutic grade proteins

Impurities	Detection techniques	Permissible limits
1. Proteinaceous (e.g. host cells and other proteins)	• SDS-PAGE electrophoresis (Coomassie brilliant blue/ silver staining) • Analytical HPLC (GPC/ RP-HPLC) • Immunoassays	< 5% No single impurity should exceed 1% of the total protein
2. DNA	• DNA dot-blot hybridization	≤ 10 pg DNA/dose
3. Pyrogenic substances (endotoxins)	• LAL • USP rabbit pyrogen test • Endogenous pyrogen test	≤ 100 EU/dose
4. Abnormal forms of the target protein	• Tryptic mapping/amino acids analysis • IEF/SDS-PAGE • RP-HPLC • GPC-HPLC • N- and C-terminus analysis • Mass spectrometry	Complete absence/lowest possible level
5. Chemical contaminants	• HPLC	Free from any chemical contaminants
6. Microbial	• Sterility test	Free from any viable microbial contaminants

both these impurities are very sensitive, as these exert their effects in the ng to pg range. Pyrogenic contaminants are most complex in nature, so three different tests are performed to probe their presence and level in the sample. The limulus amoebocyte lysate (LAL) test is only specific for endotoxin originated from the cell wall of *E. coli*. On the other hand, the USP rabbit pyrogen test covers a broad spectrum of pyrogens. A certain group of substances are non-pyrogenic to rabbit and also pass the LAL test, but when administered to humans are found to generate a pyrogenic response. These substances are detected by an endogenous pyrogen test. Human blood monocytes are cultured in vitro in presence of the sample to be tested for pyrogen. In response to the endotoxin/pyrogen present in the sample, blood monocytes release an endogenous pyrogen (IL-1), which is quantitated by RIA [159].

In order to obtain consistency in the final product in terms of impurities, each process step is validated at regular intervals. By this technique, the performance of each step is critically evaluated in terms of its ability to reduce the contaminants/risk factors. The chromatography systems are validated by spiking the initial feed material (containing target protein) with the offending substances, such as DNA, endotoxin etc. The effluent is analyzed in terms of the offending substances. QAE-anion exchange matrix in known to reduce 3 log DNA and 2 log endotoxin from the product stream (Table 6). The anion exchange step is said to be validated if similar clearances of DNA and endotoxin are obtained in such spiking tests.

The risk factors are automatically reduced if the purification is carried out as per the principles of good manufacturing practice (GMP). The major purification steps, such as ultrafiltration, desalting, and chromatography should be carried out in a clean room of class 100. Depyrogenated equipment, chromatography columns, and matrices are advisable to be used. All washings and buffer preparations should be carried out in the pyrogen-free water.

10 Case Studies

Two individual case studies are cited in this review to give an overall idea of the purification scheme of therapeutic proteins obtained from inclusion bodies.

10.1 Tissue Plasminogen Activator

Tissue plasminogen activator has been produced in an insoluble denatured form (5–10% of total stainable protein), with an average yield of 460 mg l^{-1} of spent medium [160]. The product has been isolated, refolded and purified into a functional form with an overall efficiency of 2.8%, and a purity of 99% [130]. In order to achieve 20% refolding efficiency the refolding process has been carried out from a solution of 2.5 mg l^{-1} protein concentration [160]. The details of the purification scheme are shown in Fig. 11.

10.2 Insulin

Recombinant insulin has been produced by expressing multiple joined proinsulin genes in E. coli. The proinsulin expresses and accumulates in the form of inclusion bodies as 25% of the total stainable proteins [91]. The insulin precursor contains nine amino acid residues corresponding to the first eight residues of E. coli β-galactosidase and a methionine linker, while individual proinsulin molecules are separated by a five amino acid linker ending with methionine [91]. Upon cyanogen bromide (CNBr) treatment, proinsulin monomers are quantitatively liberated. The details of the purification scheme are shown in Fig. 12. In proinsulin analog, the disulfide bonds are introduced by a thiol-mediated exchange process. About 65% refolding yield is obtained at 200 mg l^{-1} protein concentration. Refolded proinsulin has been converted to insulin upon brief reactions with trypsin and carboxypeptidase-B. Finally, the bioactive insulin molecules are separated by RP-HPLC.

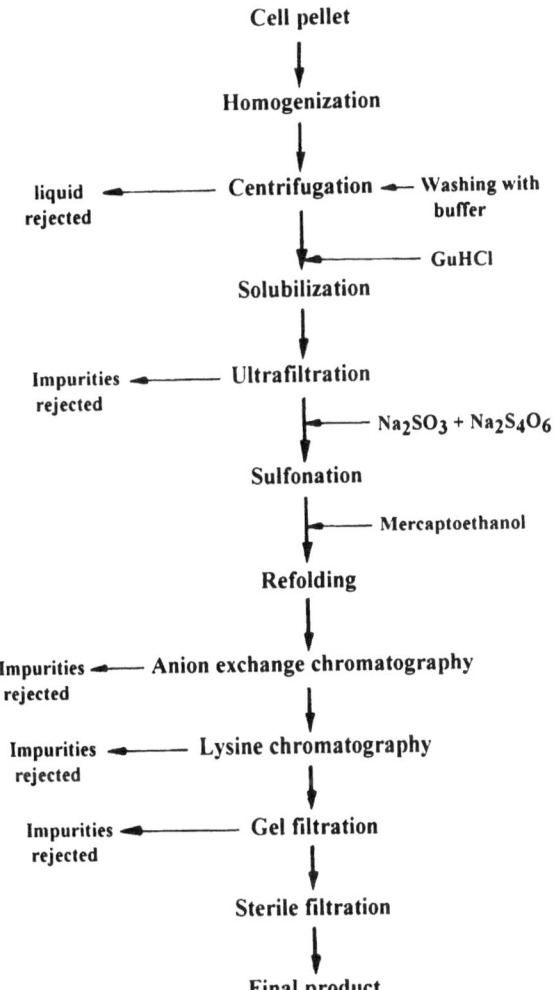

Fig. 11. Scheme for the purification of bioactive t-PA from inclusion body [130]

11 Conclusions

This review has addressed the mechanism of inclusion body formation, in vivo protein folding, alternate techniques for in vitro folding and related problems, purification, and safety aspects of therapeutic proteins. The route of inclusion body proved to be an alternative technologically and economically effective means for producing therapeutic proteins, provided glycosylation is not important for biological activity, it does not contain many disulfide bonds (maximum number may be 4), and the protein is not a hetero-oligomeric multi-subunit type.

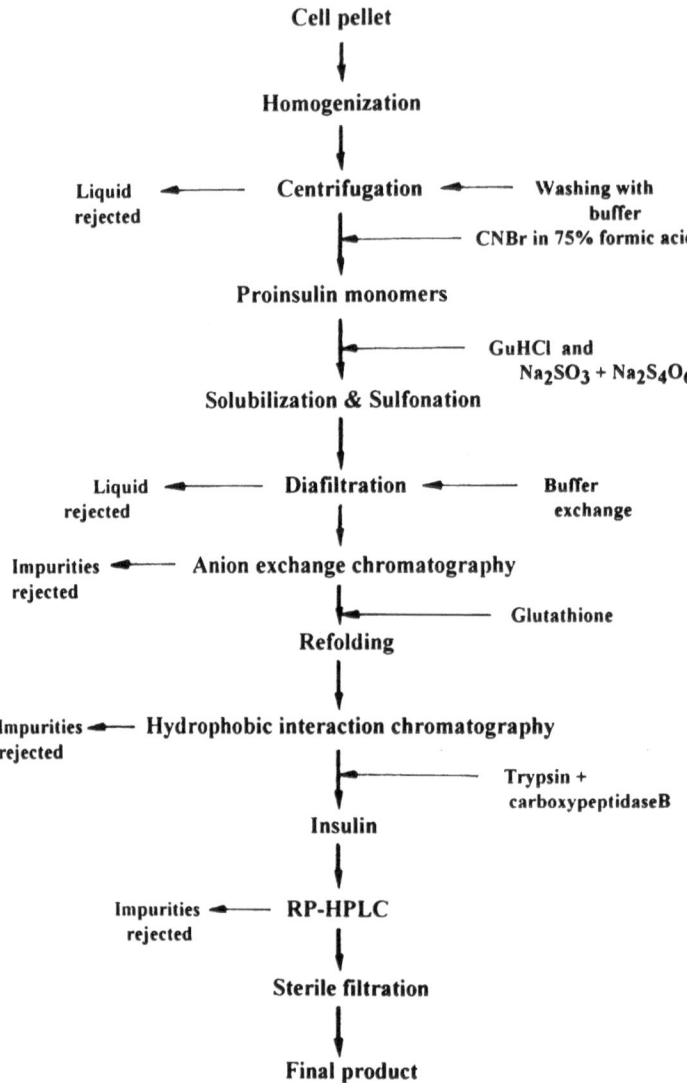

Fig. 12. Scheme for the purification of bioactive insulin from inclusion body [91]

The potential of this technology for the production of therapeutic grade protein depends on the manufacturing costs. The two main cost elements identified are material costs (88%) for purification, and equipment costs (75%) for refolding of protein [130]. The material costs may be reduced several-fold if alternative cheaper materials are used, and chromatography steps are judiciously followed so as to reduce the number of buffer exchange etc.

Refolding of protein, the most crucial step in the processing of inclusion body, needs more attention, though alternative techniques have been explored to improve yield. The problem of refolding is magnified with an increase in the number of disulfide bonds and hydrophobicity in the protein.

In order to reduce the costs of purification, it is necessary to carry out refolding from higher concentrations of protein. Two major objectives in this case will be avoiding the aggregation of protein and formation of disulfide isomers. Since the factors that determine protein folding are not completely known, no general rules and procedures have been developed for the purification of bioactive proteins in high yields from inclusion body. Chaperones and PDI are known as effective helper proteins responsible for higher folding yields in vivo, and some investigators have tried these in vitro to improve the yields. Use of helper proteins in folding in vitro may be cost effective if these are reused. One way of reusing chaperones and PDI is to immobilize them on matrices packed in a column. Protein is refolded while passing through the column, whereas helper proteins remain attached on the matrices. To date, no encouraging results have been reported on this aspect. It is also important to explore the possibility of protein refolding in fed-batch and in continuous systems. By these means the capacities of refolding equipment could be considerably minimized. In fed batch mode, since refolding is initiated from a dilute protein solution and its concentration is gradually built-up, it is possible to avoid self-association of the partially folded protein intermediates. High yield of active t-PA has been obtained from $100 \ mg \ l^{-1}$ protein concentration in a fed-batch refolding process [90]. For successful refolding of proteins in fed-batch and in continuous systems, it is of paramount importance to understand the kinetics of refolding and environmental influence on it, and the association of monomers in the case of a multi-subunit protein.

The bioactive form of a protein is labile to any change in the microenvironment, and therefore harsh physical and chemical conditions are preferably to be avoided while purifying it. Protein at a highest purity level is not only the objective of a purification scheme; it should be biologically and immunologically active also. Therefore, it is essential to restore the functional properties of a protein while purifying it, and at the same time attempts should be made to maximize the yield and to reduce the load of contaminants.

Acknowledgements. The author is grateful to his colleagues Drs. Prakash Bhatia, R. P. Roy, Amulya K. Panda. Deepak K. Giri, Lalit C. Grag and Mr. Suresh Chandran for useful suggestions and for critically reviewing the articles.

12 References

1. Hansen L, Blue Y, Barone K, Collen D, Larsen GR (1988) J Biol Chem 263: 15713
2. Anon (1992) Eur Biotech Newsletter 139: 1
3. Schumacher G, Sizmann D, Huag H, Bock A (1986) Nucleic Acids 14: 5713

4. Nishimura N, Komatsubara S, Taniguchi T, Kisumi M (1987) J Biotechnol 6: 31
5. Lowe PA, Rhind SK, Surgue R, Marston FAO (1987) Solubilization, refolding and purification of eukaryotic proteins expressed in *E. coli*. In: Burgess R (ed) Protein purification: micro to macro. Alan R Liss, New York, p 429
6. Marston FAO (1986) Biochem J 240: 1
7. Cleland JL (1993) Impact of protein folding on biotechnology. In: Cleland JL (ed) Protein Folding – *In vivo* and In *vitro*. ACS Symposium Series 526: 1
8. Wittrup KD, Mann MB, Fenton DM, Tsai LB, Bailey JE (1988) Bio/Technol 6: 423
9. William DC, van Frank RM, Muth WL, Burnett JP (1982) Science 215: 687
10. Biotechnology Medicines in Development – Genetic Engineering News (1992) Jan, p. 27
11. Office of Technology Assessment (1991) In: Brown B (ed) Biotechnology in a Global Economy. OTA, US Congress Report No. OTA-BA-494, Washington, DC, US Government Printing Office.
12. Schein CH (1989) Bio/Technol 7: 1141
13. Krueger JK, Kuke MN, Schutt C, Stock O (1989) BioPharm 2: 40
14. Mitraki A, King J (1989) Bio/Technol 7: 690
15. Parente D, Deferra F, Galli G, Gandi G (1991) FEMS Lett 77: 243
16. Cousens, L, Shuster JR, Gallegos C, Ku L, Stempien MM, Urdea MS, Sanchez-Pescardor R, Taylor A, Tekamp-Olsen P (1987) Gene 61: 265
17. Thomas CP, Booth TF, Roy P (1990) J Gen Virol 71: 2073
18. Sedir JM (1988) Bio/Technol 6: 1192
19. Modrich P, Zabel D (1976) J Biol Chem 251: 5866
20. Botterman J, Zabean M (1985) Gene 7: 229
21. Kane JF, Hartley DL (1991) Properties of recombinant protein containing inclusion bodies in *E. coli*. In: Seetharam R, Sharma SK (eds) Purification and Analysis of Recombinant Proteins. Marcel Dekker Inc. New York, p 121
22. Schoemaker JM, Brasnett AH, Marston FAO (1985) EMBO J 4: 775
23. Mizukami T, Komatsu Y, Hosoi N, Itoh S, Oka T (1986) Biotechnol Lett 8: 605
24. Goldenberg D, King J (1982) Proc Natl Acad Sci, USA 5: 277
25. Stewart DE, Sarkar A, Wampler JE (1990) J Mol Biol 214: 253
26. Kane JF, Hartley DL (1988) Trends in Biotechnol 6: 95
27. Wulfing C, Plückthum A (1994) Mol Microbiol 12: 685
28. Georgious G, Telford JN, Shuler ML, Wilson DB (1986) Appl and Environ Microbiol 52: 1157
29. Ellis J (1987) Nature 328: 378
30. Baneyx F, Gatenby AA (1993) GroEL-mediated Protein Folding. In: Cleland JL (ed) Protein Folding – *In Vivo* and In *Vitro*. ACS Symposium Series. American Chemical Society, Washington DC, 526: 133
31. Kenealy WR, Gray JE, Ivanoff LA, Tribe DE, Reed DL, Korant BD, Petteway SR Jr. (1987) Dev Ind Microbiol 28: 45
32. Bowden GA, Paredes AM, Georgiou G (1991) Bio/Technol 9: 725
33. Taylor G, Hoare M, Gary DR, Marston FAO (1986) Bio/Technol 4: 553
34. Rose GD, Geselowitz AR, Lesser G-J, Lee RH, Zehfus MH (1985) Science 229: 834
35. Havel HA, Kauffman EW, Plaisted SM, Brems DN (1986) Biochemistry 25: 6533
36. Kyte J, Doolittle RF (1982) J Mol Biol 157: 105
37. Tanford C (1961) Physical Chemistry of Macromolecules. John Wiley & Sons, New York, p 241
38. Wilkinson DL, Harrison RG (1991) Bio/Technol 9: 443
39. Freeman RB (1992) Protein folding in the cell. In: Creighton TE (ed) Protein Folding. WH Freeman & Company, New York, p 455
40. Hlodan R, Hartl FU (1994) How the protein folds in the cell. In: Pain RH (ed) Mechanisms of Protein Folding. IRL Press, Oxford, p 194
41. Gilbert HF (1990) Adv Enzymol 63: 69
42. Randall LL, Hardy SJS (1989) Science 243: 1156
43. Dermann AI, Puziss JW, Bussford PJ, Beckwith J (1993) EMBO J 12: 879
44. Bychkova VE, Pain RH, Ptitsyn OB (1988) FEBS Lett 238: 231
45. Weiss JR, Roy PH, Bassford PJ Jr. (1988) Proc Natl Acad Sci, USA 85: 8978
46. Zahn R, Plückthum A (1992) Biochemistry 31: 3250
47. Wickner W (1984) Trends Biochem Sci 14: 280
48. Georgopoulos C (1992) TIBS 17: 295

49. Jacq A, Holland B (1993) Curr Op Struct Biol 3: 541
50. Matsuyama S-I, Fujita Y, Mizushima S (1993) EMBO J 12: 265
51. Wickner W, Drissen AJM, Harti F-V (1991) Annu Rev Biochem 60: 101
52. Bardwell JCA, McGrovern K, Beckwith J (1991) Cell 67: 581
53. Akiyama Y, Ito K (1993) J Biol Chem 168: 8146
54. Knappik A, Krebber C, Plückthum A (1993) Bio/Technol 11: 77
55. Bardwell JCA, Lee J-O, Jander G, Martin N, Belin D, Beckwith J (1993) Proc Natl Acad Sci, USA, 90: 1038
56. Lin J, Walsh CT (1990) Proc Natl Acad Sci, USA 87: 4028
57. Tanford C (1968) Protein denaturation. Adv Protein Chem 23: 121
58. Scholtz JM, Baldwin RL (1992) Annu Rev Biophys Biomol Struct 21: 95
59. van Mierlo CPM, Darby NJ, Creigton TE (1992) Proc Natl Acad Sci, USA, 89: 6775
60. Matouschek A, Serrano L, Fersht AR (1992) J Mol Biol 224: 819
61. Pace CN, Grimsley GR, Thomson JA, Barnett BJ (1988) J Biol Chem 263: 11820
62. Karplus M, Shakhnovich E (1992) Protein Folding: Theoretical studies of Thermodynamics and Dynamics. In: Creighton TE (ed.) Protein Folding. WH Freeman and Company, New York, p. 127
63. Creighton TE (1994) The protein folding problem. In: Pain RH (ed) Mechanisms of Protein Folding. IRL Press, Oxford, p. 1
64. Jaenicke R (1991) Biochemistry 30: 3147
65. Anfinsen CB (1973) Science 181: 223
66. Fischer G, Schmid FX (1990) Biochemistry 29: 2205
67. Brandts JF, Halvorson HR, Brennan M (1975) Biochemistry 4: 4453
68. Nall BT (1994) Proline isomerization as a rate-limting step. In: Pain RH (ed) Mechanism of Protein Folding, IRL Press, Oxford, p. 80
69. Fischer G, Bang H, Mech C (1984) Biomed Biochim Acta 10: 1101
70. Lang K, Schmid FX, Fischer G (1987) Nature 329: 268
71. Schonbrunner ER, Schmid FX (1992). Proc Natl Acad Sci, USA, 89: 4150
72. Anfinsen CB and Scheraga HA (1975) Adv Protein Chem 291: 205
73. Jaenicke R, Rudolph R (1992) Protein Folding. In: Creighton TE (ed) Protein Structure, IRL Press, Oxford, p. 191
74. Noiva R, Lennarz WJ (1992) J Biol Chem 267: 3553
75. Creighton TE (1992) Folding pathways elucidated using disulfide bonds. In Creighton TE (ed) Protein Folding. WH Freeman and Company, New York p. 301
76. Konishi Y, Ooi T, Scherage HA (1981) Biochemistry 20: 3445
77. Gilbert HF (1994) The formation of native disulfide bonds. In: Pain RH (ed) Mechanisms of Protein Folding. IRL Press, Oxford, p. 104
78. Poland DC, Scheraga HA (1965) Biopolymers 3: 379
79. Oslon KC (1985) US Patent 4518526
80. Lyles MM, Gilbert HF (1991) Biochemistry 30: 613
81. Saxena UP, Wetlanfer DB (1970) Biochemistry 9: 5015
82. Buchner J, Rudolph R (1991) Bio/Technol 9: 157
83. Mukhopadhyay A, Paul R, Batra JK (1995) Biochem J (submitted)
84. London J, Skrzynia C, Goldberg M (1974). Eur J Biochem 47: 409
85. Rudolph R. Zettlmeissl G, Jaenicke R (1979) Biochemistry 18: 5572
86. Zettlemeissl G, Rudolph R, Jaenicke R (1979) Biochemistry 18: 5567
87. Mitraki A, Betton J-M, Desmadril M, Yon J (1987) Eur J Biochem 163: 29
88. Brems DN (1988) Biochemistry 27: 4541
89. Halenbeck R, Kawasaki E, Wrin J, Koths K (1989) Bio/Technol 7: 710
90. Rudolph R, Fischer S, Matters R (1987) International Patent Application WO 87/02673
91. Cockle S, Lennick M, Shen S-H (1987) Production of peptide hormones in E. coli via multiple joined genes. In: Burgess R (ed) Protein Purification-Micro to Macro. Alan R. Liss Inc., New York, p. 375
92. Cleland JL, Wang DIC (1991) Equilibrium association of a molten globule intermediate in the refolding of bovine carbonin anhydrase. In: Georgiou G, Bernardez-Clark, ED (eds) Protein Folding. ACS Symposium Series 470: 169
93. Zaheer A, Lim R (1991) Oxidative refolding of recombinant human glia maturation factor beta. ibid, p. 79

94. Lu HS, Clogston CL, Merewether LA, Narhi LO, Boone TC (1993) Role of disulfide bonds in folding of recombinant human granulocyte colony stimulating factor produced in *E. coli* In: Cleland JL (ed) Protein Folding *In vivo* and *In vitro*. ACS Symposium series 526: 184
95. Weir MP, Sparks J (1987) Biochem J 245: 85
96. Hopkins TR (1991) Physical and chemical cell disruption for the recovery of intracellular proteins. In: Seetharam R, Sharma SK (eds) Purification and Analysis of Recombinant proteins. Mercel Dekker Inc., New York, p. 57
97. Mizushima S (1986) Proc Bio Fair, Tokyo, p 262
98. Lilly MD, Dunnill P (1969) Isolation of intracellular enzymes from microorganisms: The development of a continous process. In: Perlman D (ed) Fermentation Advances, Academic Press, New York, p. 225
99. Nerc HC, Heppel LA (1964) Proc Natl Acad Sci, USA, 51: 1267
100. Hetter D, Wang H (1989) Biotechnol Bioeng 33: 886
101. Engler CR (1990) Cell disruption by homogenization. In: Asenjo JA (ed) Separation Processes in Biotechnology. Marcel Dekker Inc., New York, p. 95
102. Agerkvist I, Enfors S-O (1990). Biotechnol Bioeng 36: 1083
103. Middelberg APJ, O'Nell BK, Bogle IDL, Snoswell MA (1991) Biotechnol Bioeng 38: 363
104. Babbitt PC, West BL, Buechter DD, Kuntz ID, Kenyon GL (1990) Bio/Technol 8: 945
105. McCaman MT (1992) US Patent 5082775
106. Koths K, Thompson J, Kunitani M, Wilson K, Hanisch W (1986) US Patent 4569790
107. Rausch SK, Merg H (1985). World Patent 87/102800/15
108. Evans TW, Knuth MW (1986) Eur Patent 0263902.
109. Korant BD (1987) Eur Patent 0210846
110. Kronheim SR, Cantrell MA, Deely MC, March CJ, Glackin PJ, Anderson DM, Hemenway T, Merriam JE, Cosman D, Hoppe TP (1986) Bio/Technol 4: 1079
111. McCoy KM, Frozt RA (1990). Eur Patent 0373325
112. Lowe PA, Rhind SK, Sugrue R, Marston FAO (1987) Solubilization, refolding and purification of eukaryotic proteins expressed in *E. coli*. In: Burgess R (ed) Protein Purification - Micro to Macro. Alan R. Liss Inc., New York, UCLA Symposium Series 68: 429
113. Hart RA, Bailey JE (1992) Biotechnol Bioeng 39: 1112
114. Sharific BG, Bascom CC, Khurana VK, Johnson TC (1985) J Chromatogr 324: 173
115. Montelaro RC, West M, Issel CJ (1981) Anal Biochem 114: 398
116. Welling GN, Groen G, Welling-Wester S (1983) J Chromatogr 266: 629
117. Lu RC, Elzinga M (1977) Anal Biochem 77: 243
118. Hermann R (1993) Protein Folding. European Patent Office, Netherlands, EPO Applied Technology Serises, v. 12
119. Weir MP, Sparks J, Chaplin AM (1987) J Chromatogr 396: 209
120. Langley KE, Berg TF, Strickland TW, Fenton DM, Boone TC, Wypych J (1987) Eur J Biochem 163: 313
121. Orsini G, Brandazza A, Sarmientos P, Molinari A, Lamsen J, Canet S (1991) Eur J Biochem 195: 691
122. Lu HS, Clogston CL, Narhi LO, Merewetter LA, Pearl WR, Boone TC (1992) J Biol Chem 267: 8770
123. Thatcher DR, Hitchock A (1993) Protein Folding in Biotechnology In: Pain RH (ed) Mechanism of Protein Folding, IRL Press, Oxford, p. 229
124. Wilhelm OG, Jaskunas SR, Vlahos CJ, Bang NV (1990) J Biol Chem 265: 14606
125. Yamunichi K, Takahashi M, Nishida T, Ohmoto Y, Takano M, Nakai S, Hirai Y (1991) J Biochem 109: 404
126. Jones AJS, Olson KC, Shire SJ (1985) US Patent 4512922
127. Builder SE, Ogez JR (1986) US Patent 4620948
128. Orsini G and Golberg ME (1978) J Bio Chem 253: 3453
129. Creighton TE (1979) J Mol Biol 129: 235
130. Datar RV, Cartwright T, Rosen C-G (1993) Bio/Technology 11: 349
131. Winkler ME, Blaber M (1986) Biochemistry 25: 4041
132. Jaenicke R, Rudolph R (1989) Folding Protein. In: Creighton TE (ed) Protein Structure – A Practical Approach. IRL Press, Oxford, p. 191
133. Burnier JP, Johnston PD (1988) Eur Patent 0251615
134. Valax P, Georgious G (1991) Folding and Aggergation of RTEM β-lactamase. In: Georgious G, De Bernardez-Clark E (eds) ACS Symposium Series, American Chemical Society, Washington DC, 470: 97

135. Sawano H, Kuomoto Y, Ohta K, Sasaki Y, Segawa S, Tachibana H (1992) FEBS Lett 303: 11
136. Cleland JL, Builder SE, Swartz JE, Winkler M, Chang JY, Wang DIC (1992) Biotechnol 10: 1013
137. Marchiani DJ, Hung C-H, Cheng K-L, Kensil C (1987) Solubilization of inclusion body proteins by reversible N-acylation. In: Burgess R (ed) Protein Purification-Micro to Macro. UCLA Symposium Series, Alan R. Liss Inc., New York; 68: 443
138. Hagen AJ, Hatton TA, Wang DIC (1990) Biotechnol Bioeng 35: 955
139. Bonnerjea J, Oh S, Hoare M, Dunnill P (1986) Biotechnol 4: 954
140. Staehelin T, Hobbs DC, Kung H, Lai C-Y, Pestka S (1981) J Biol Chem 256: 9750
141. Hochuli E (1986) Chimia 40: 408
142. Garg VK, Costello MAC, Czuba BA (1991) Purification and production of therapeutic grade proteins. In: Seetharam R, Sharma SK (eds) Purification and Analysis of Recombinant Proteins. Marcel Dekkar Inc., New York, p. 29
143. Kniep EM, Kniep B, Grote W, Contradt HS, Monner DA, Mühlradt P (1984) Eur J Biochem 143: 199
144. Johannsen HS, Tan YH (1983) J Interferon Res 3: 473
145. Low D (1990) Scale up of protein chromatography separations. In: Hancock WS (ed) High Performance Liquid Chromatography in Biotechnology. John Wiley & Sons, New York, p. 117
146. Ben-Bassat A (1991) Method for removing N-terminal methionine from recombinant protein. In: Seetharam R, Sharma SK (eds) Purification and Analysis of recombinant proteins. Marcel Dekker Inc., New York, p. 147
147. Glasbrenner K (1986) JAMA 255: 581
148. Ben-Basser A, Bauer K, Chang SY, Myambo Y, Boosman A, Chang S (1987) J Bacteriol 169: 751
149. Flinta C, Persson B, Jornvall H, von Heijne G (1986) Eur J Biochem 154: 193
150. Christensen T, Dalboge H, Snel L (1991) Post-biosynthesis modification – human growth hormone and insulin precursors. In: Chin Yuan-Yuan H, Gueriguian JL (eds) Drug Biotechnology Regulation. Mercel Dekker Inc., New York, p. 206
151. Garnick RL, Ross MJ, Baffi RA (1991) Characterization of proteins from recombinant DNA manufactured. ibid. p. 263
152. Hearn JP, Gidley-Baird AA, Hodges JK, Summers PM, Wibley GE (1988) J Reprod Fertil (supplement) 36: 49
153. Weickmann JL, Elson ML, Glitz DG (1981) Biochemistry 20: 1272
154. Lange B, Valtieri M, Santoli D, Caracciolo D, Mavilio F, Gemperlein I, Griffin C, Emannel B, Finan J, Nowell P, Rovera G (1987) Blood 70: 192
155. Dighe RR, Moudgal NR (1983) Biochem Biophys 225: 490
156. Garnick RL, Ross MJ, du M'ee CP (1988) Analysis of recombinant biologicals. In: Swarbrick J, Boylan JC (eds) Encyclopedia of Pharmaceutical Technology. Marcel Dekkar, New York 1: 253
157. Garnick RL (1988) The purity of biotechnology products from an industrial perspective. Presented at the 2nd Annual Seminar on Analytical Biotechnology (May 7-20), Sheraton Inner Harbor Hotel, Baltimore, Barr Enterprises, Walkersville, MD.
158. Harada Y (1987) Problems presented by animal toxicity studies. In: Graham CE (ed) Progress in Clinical and Biological Research - Preclinical Safety of Biotechnology Products Intended for Human Use. Alan R. Liss Inc., New Yrok, 235: 127
159. Dinarello CA, Renfer L, Wolff SM (1977) Proc Natl Acad Sci USA, 74: 4624
160. Sarmientos P, Duchesne M, Denefle P, Boiziau J, Fromage N, Delporte N, Parker F, Lelievre Y, Mayaux JF, Catwright T (1989) Biotechnol 7: 495

Extraction and Purification of Cephalosporin Antibiotics

A.C. Ghosh, R.K. Mathur and N.N. Dutta
Regional Research Laboratory, Jorhat 785 006, India

1 Introduction 112
2 Product Distribution and Assay of Spent Medium 115
  2.1 Cell Mass Separation and Isolation 115
  2.2 Microbiological Assay 117
3 Methods of Extraction 118
  3.1 Solid Phase Adsorption 118
  3.2 Liquid-Liquid Extraction 120
    3.2.1 Extraction via Lipophilic Intermediates 120
    3.2.2 Extractive Esterification 121
    3.2.3 Reactive Extraction 121
  3.3 Liquid Membrane Extraction 122
    3.3.1 Emulsion Liquid Membrane 122
    3.3.2 Supported Liquid Membrane 124
    3.3.3 Non-Dispersive Extraction in Hollow Fiber Membrane 124
    3.3.4 Emerging Liquid Membrane Processes 125
  3.4 Membrane (Synthetic) Separation: UF/MF/RO/ED 125
  3.5 Aqueous Two-Phase Partitioning 126
4 Methods of Purification 127
  4.1 Selective Precipitation and Crystallization 127
  4.2 Separation and Purification by Chromatography 128
    4.2.1 Thin layer and Paper Chromatography 128
    4.2.2 Ion-Exchange Chromatography 129
    4.2.3 Hydrophobic Interaction and Hydroxyapatite Chromatography 130
    4.2.4 Gel Filtration/Size Exclusion Chromatography 134
    4.2.5 High Performance Liquid Chromatography 134
  4.3 Electrophoretic Techniques 135
    4.3.1 Polyacrylamide Gel Electrophoresis 139
    4.3.2 Isoelectric Focussing 139
5 Process Design and Scale-Up Consideration 140
6 References 142

The biologically active natural and semisynthetic cephalosporin antibiotics require proper methods of extraction and purification for their isolation and subsequent pharmacological studies. This article reviews the various methods useful for extraction and purification of individual compounds as well as the enzymes involved in their biosynthesis. Applicability of the methods for downstream processing of the spent medium has been critically analysed. Adsorption chromatography, particularly with reverse phase materials, in combination with membrane separation is the most successful technique for extraction as well as purification of most of the enzymes and individual compounds. Techniques such as reactive extraction in liquid membrane, non-dispersive extraction in hollow fiber membrane and aqueous two-phase extraction are likely to emerge in new generation processes. Finally, some aspects of process design and scale-up have been discussed, highlighting the research needs of pragmatic importance.

Advances in Biochemical Engineering/
Biotechnology, Vol. 56
Managing Editor: Th. Scheper
© Springer-Verlag Berlin Heidelberg 1997

1 Introduction

The cephalosporin group of β-lactam antibiotics constitutes a large portion of the multibillion dollar antibiotics market. Since discovery of the Cephalosporin-C (CPC) molecule in 1955 [1], the processes for production of natural and semisynthetic cephalosporins have undergone various modifications. Though the antibiotics are produced by a limited group of microorganisms (bacteria and fungi), the biosynthetic pathway (Fig. 1) for their production is now fairly well understood [2]. Consequent to the development of high yielding strain of *Cephalosporium acremonium*, understanding of the enzymes involved in the biosynthesis has been considerably improved. There are about 13 therapeutically important semisynthetic cephalosporins, but only a limited species of natural cephalosporins are known so far (Tables 1 and 2). In most instances, the 7-α-position on the β-lactam ring has a delta linked D-α-aminoadipyl side chain (R′), but three cephalosporin compounds have been identified with glutaryl side chains and one with *N*-acetyl derivatives of cephalosporins.

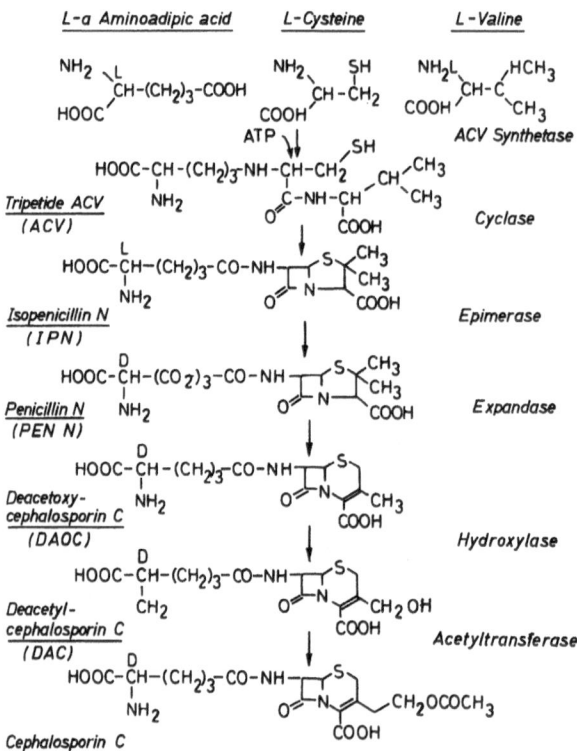

Fig. 1. The biosynthetic pathway of cephalosporin-C from the amino acids, L-α-aminoadipic L-cysteine L-valine

Table 1. Naturally produced cephalosporins

Compound	R^1	R^2	R^3	Origin
Deacetoxy cephalosporin C	D-α Amino adipic acid	–H	–H	Fungi, streptomycetes, bacteria
Deacetyl cephalosporin C	–do–	–do–	–OH	Fungi, streptomycetes
Cephalosporin-C	–do–	–do–	–OCOCH₃	C. acremonium
F-1	–do–	–do–	–SCH₃	–do–
C-1778a	Glutaric acid	–do–	–H	–do–
C-1778b	–do–	–do–	–OH	–do–
C-1778c	–do–	–do–	–O COCH₃	–do–
C-43-219	D-α Amino adipic acid	–do–	$-S-\overset{H_3C}{\underset{H_3C}{C}}-\overset{COOH}{\underset{NH_2}{CH}}$	–do–
N-Acetyl deacetoxy cephalosporin-C	N-Acetyl derivative of D-α Amino adipic acid	–do–	–H	–do–
A16886 A	D-α Amino adipic acid	–do–	–OCONH₂	–do–

Nutritional requirements of *C. acremonium* are a number of sugars, methyl oleate, or glycerol as carbon source, and inorganic nitrogen, amino acid or complex polypeptide as nitrogen sources. The basic studies on biosynthesis of cephalosporins have progressed very slowly for many years, but recent developments in molecular biology techniques have provided new impetus [3]. Genetic aspects have been studied, mainly in relation to CPC production. Use of recombinant DNA technique has been reported to increase the production by about 15% in an industrial strain of *C. acremonium* [4]. Gene cloning from both fungal and actinomycetes sources has been achieved at least with partial success [5]

Microbial cells and their enzymes like penicillin-G acylase and CPC acylase [6] can produce β-lactum nuclei of 7-aminodeacetoxycephalosporanic acid (7-ADCA) and 7-aminocephalosporanic acid (7-ACA) which are in turn used for the production of semi-synthetic cephalosporins. Penicillin-G acylases from *Escherichia coli*, *Bacillus magaterium* and *Penicillium ruttqgeri* and Penicillin V acylase as well as intact cells of *Erwinia aroidae* hydrolyse 7-phenyl acetoamido 7-ADCA into 7-ADCA [7] Acylase enzymes from *Acetobacter turbidians*, *Xanthomonas citri*, and *Achromobacter* species can produce cephalexin from 7-ADCA [8].

114   A.C. Ghosh, R.K. Mathur and N.N. Dutta

Table 2. Semisynthetic cephalosporins

Compound	R^1	R^2	R^3
Cephalexin	phenyl–CH(NH$_2$)–CO	–H	–H
Cephalothin	thienyl–CH$_2$–CO	–H	$-O-CO-CH_3$
Cephaloglycin	phenyl–CH(NH$_2$)–CO	–H	$-O-CO-CH_3$
Cephradin	cyclohexadienyl–CH(NH$_2$)–CO	–H	–H
Cephaloridine	thienyl–CH$_2$–CO	–H	$-N^+$ pyridinium
Cefazolin	tetrazolyl–CH$_2$–CO	–H	$-S-C(N-N)=CH_3$ thiadiazole
Cefuroxime	furyl–C(N–OCH$_3$)–CO	–H	$-OCONH_2$
Cefoxitin	thienyl–CH$_2$–CO	$-OCH_3$	$-OCONH_2$
Cefataxime	aminothiazolyl–C(N–OCH$_3$)–CO	–H	$-OCOCH_3$

7-ACA, the intermediate for cephaloglycine, cephalothin, cephapirin, cephazolin, and cephamandole is usually obtained by complex chemical removal of the natural D-α-aminoadipic acid side chain of CPC. Direct enzymatic hydrolysis of CPC into 7-ACA with true CPC acylase of *Pseudomonas* strains is possible [9]. However, only few enzymes capable of direct conversion have been discussed, probably because of the unusual nature of the D-aminoadipyl side chain of CPC and very low specific activity of these enzymes [10]. A two-step enzymatic process may be more attractive. In this process, CPC is oxidised and deaminated by intracellular D-amino acid oxidase of *Trigonopsis variabilis* or *Rhodotorula gracilis* [11, 12] into 5-carboxy-5-oxopentane-7-aminocephalosporanic acid which, upon decarboxylation, spontaneously yields glutaryl (GL) 7-ACA for 7-α-(4-carboxbutane amino)-cephalosporanic acid. The GL 7-ACA can be hydrolysed by a specific GL 7-ACA acylase [13]. Cloning of genes for acylases of *Pseudomonas* species in *E. coli* has been examined for industrial application [14]. Enzymatic deacetylation of cephalosporins can produce cefuroxime and S-1108 [15]. CPC deacetylase or acetylhydrolase (CAH) activity has been found in various sources such as citrus peel, actinomycetes, and fungi, as well as *Bacillus subtilis*, but the latter source provides as enzyme which converts CPC to deacetylcephalosporin with high efficacy [16].

During biochemical synthesis, the product distribution can vary depending on the source of microorganism as well as the culture conditions. Detection of biological activity and quantification of the products in the spent medium are of paramount importance. Efficient extraction and purification methods are necessary, not only for analytical purposes but also for large scale application. A total developmental programme for β-lactam production involves identification and quantification of products with an assay of their biological activity and a search for an efficient extraction and purification procedure. This article provides an overview of the procedures used for extraction and purification of cephalosporin antibiotics on an industrial scale as well as the techniques used for enzymes involved in their biosynthesis.

2 Product Distribution and Assay of Spent Medium

2.1 Cell Mass Separation and Isolation

The separation of cell biomass is the primary recovery operation in a downstream processing train (Fig. 2). The synthesis of natural cephalosporins has been proved mostly in cell free systems [2] and the product of secondary metabolism is excreted into the media containing complex ingredients. The largest production of naturally occurring CPC (200 tonnes/annum) is by the

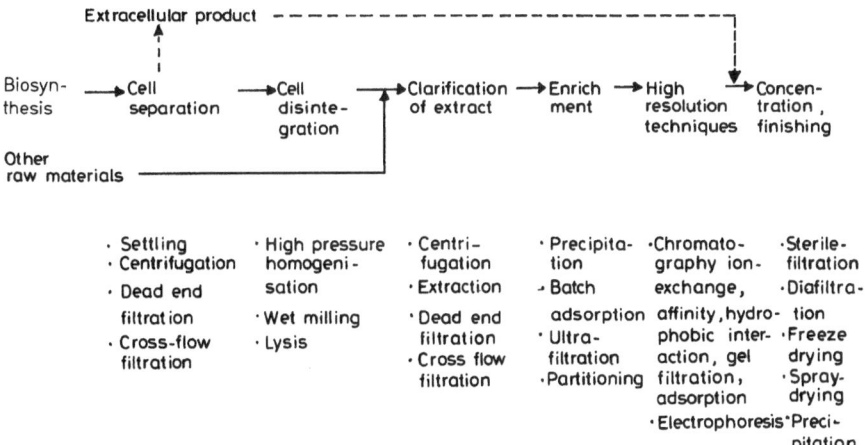

Fig. 2. Downstream processing of fermentation broth

eukaryote (true fungi), *C. acremonium*. The filtration properties of the spent medium, however, are essentially the same irrespective of the type of microorganism used. The wet mass content in the medium of *C. acremonium* is higher than 25 vol.% with concomitant by-product deacetyl CPC at high concentration (about 26 – 50% relative to CPC). The cells usually exhibit changes in the morphology and the product formation is associated with marked cell fragmentation leading to an increase in the viscosity of the spent medium [17]. In view of this, dilution of the spent medium is essential prior to filtration of the mycelial mass and suspended solid. Suspended solids can be removed by the methods based on gravitational, surface, and electrical phenomena, but the gravitational methods like classification/settling, centrifugation, and flocculation are more commonly practised. Laboratory filtration techniques involve ordinary filter paper and ultrafiltration membranes (50 000–10 000 MWCO). Ultrafiltration, microfiltration, and reverse osmosis membranes of polysulfone, PTFE, and cellulose acetate have been described for cell mass separation of *C. acremonium* and *S. clavuligerus* media [18–21]. Cross-flow mode of membrane filtration has been emphasised for efficient recovery of cell biomass [22]. Laboratory centrifuges equipped with continuous flow rotors can be used to obtain clarified supernatant. On an industrial scale, filtration can be carried out in a plate and frame, West Falia centrifugal separator or rotary vacuum filter using coarse cloth. The selection of a refrigerated centrifuge for process application depends on cost, flexibility, and ease of continuous operation at a desired capacity. Various types of filtration equipments and their design principles have been discussed in the literature [23]. It may be noted that genetically manipulated organisms, though producing high levels of secondary metabolites, do not impose limitations on filterability of the mycelia.

During biosynthesis of CPC, the formation of the synthesizing enzyme is sequentially induced in the metabolic pathway (Fig. 1). It is therefore necessary to separate/isolate the enzymes, particularly for analytical purposes and for understanding the mechanisms of biosynthesis. They are usually isolated as cell-free crude extract by preliminary recovery methods involving cell disruption by sonication, and harvested by centrifugation and cross-flow filtration. Mechanical as well as non-mechanical techniques are applied for cell disruption [24]. High pressure homogenisation, bead milling and ultrasonication are the most widely practised mechanical methods. Non-mechanical methods may be chemical or biochemical in nature. Cells may be permeabilised by organic solvents or by enzymatic lysis of the cell walls. The use of organic solvents implies appropriate safety precautions whereas the enzymatic processes are expensive on a large scale. By means of genetic manipulation, it becomes possible to control cell lysis and change the properties of cells in order to facilitate biomass recovery [25].

Treatment of the spent medium with osmotic lysis buffer in presence of stabilising agents is a common practice for isolating the crude cell-free extract. 4-Morpholine propane sulfonate (MOPS)-NaOH and tris-HCl-KCl are used as the buffer with 2-mercaptoethanol or dithiothreitol (DTT), glycerol, and phenyl methyl sulfonyl fluoride (PMSF) as the stabilizing agents. The enzymes so isolated are ACVS from *C. acremonium* [26, 27], *S. clavuligerus* [28, 29], *A. nidulans* [30]; isopenicillin *N* synthase from *C. acremonium* [31, 32], *Acremonium chrysogenun* [33], *S. clavuligerus/lipmani* [34], *S. lactamduran* [35, 36]; isopenicillin *N* epimerase from *C. acremonium* [36, 37], *S. clavuligerus* [38], *N. lactamduran* [39]; deacetoxy CPC synthase from *C. acremonium* [40–42], *S. clavuligerus* [42–44], *S. lactamdurans* [42, 45] and deacetyl CPC acytransferase from *C. acremonium* [46]. Almost similar techniques have been reported for other enzymes such as α-aminoacid ester hydrolase from *Xanthomonas citri* [7], *Acetobacter turbidans* [8]; CPC acylase and glutaryl 7-ACA acylase from *pseudomonas sp.* [10, 13, 14], and CPC deacetylase from *Bacillus subtilis* [16]. Use of freeze solvent method has been reported to enhance release of the GL 7-ACA acylase from the cell mass of *Pseudomonas sp.* so as to achieve more than 60% yield with high activity of the enzyme [47]. In the case of mutant strains of *X. citri*, cell disruption by sonication is not adequate to release more than 50% of the membrane-bound enzyme, yield and activity of which could be improved by addition of surfactants like Triton X-100 and sodium deoxycholate [7]. Use of cetyltrimethyl ammonium chloride has also been reported to enhance the yield of CPC acylase [10]. The whole cells of *Triginopsis variabilis* for enzymatic oxidation of CPC to GL 7-ACA could be harvested by concentration on a 0–2 μM ceramic filter followed by water washing and cells permeabilisation via treatment with either 25% acetone/water or 2% methylene chloride/water [11].

2.2 Microbiological Assay

Assay by microbiological method is rather a common practice in determining the low levels of antimicrobial activity in crude fermentation broth. Procedures for microbiological assay are well developed and described in standard text [48]. Agar diffusion methods have been developed for various semisynthetic cephalosporins like cefadroxil, cefazolin, cephalexin, cephaloglycin, cephradine, cephalothin, cephapirin and cefuroxime using *Staphylococcus aureus* ATCC 6538 P as the test organism [48]. CPC and deacetyl CPC in spent medium were determined with *Alcaligenes facalis* IFD 13111 and *Commemonas ferrigena* HO 12685 [13, 17]. Cephalosporin-synthesising enzymes and concomitant product formation were assayed using *Bacillus subtilis* NCIB 8993 and *E. coli* [28, 29, 43, 45], *Micrococcus luteus* [27, 32, 36, 39], *Staphylococcus aureus* [26, 31] and *Sarcina lutea* [31]. Turbidimetric and/or photometric assays are performed by adding test solution to liquid media previously incubated with the test organism. The basic design methodology of the assay has been described in the literature [48]. Limited information is available on the use of photometric methods for cephalosporin antibiotics and related enzymes. A dye-binding photometric method using Commassie Blue G-250 has been very frequently used to determine the protein content of spent media [13, 28]. A simple colorimetric method can be used for the determination of cephalexin content during enzymatic synthesis. This method is based on characteristic color reaction of cephalexin having absorbency at 470 nm by the addition of alkali [8]. The activity of CPC deacetylase has been determined spectrophotometrically in a system of sodium phosphate buffer and p-nitrophenyl acetate as the substrate [16].

Due to the rapid development in high performance liquid chromatography of various forms, the classical microbiological techniques now seem to be less preferred in many laboratories. However, the microbiological assay remains the most suited technique for assessing the antimicrobial activity.

3 Methods of Extraction

3.1 Solid Phase Adsorption

Adsorptive separation in principle can be effective for isolation of lipophilic and hydrophilic microbial products [49]. However, for the isolation of hydrophilic cephalosporins, only a few adsorbents (Table 3) are suitable because of the capacity limitation. These adsorbents separate the products on the basis of their differences in polarity and surface binding by Van der Waal's forces. The earlier generation of adsorbents, i.e., activated carbon, molecular sieves, silica gel and cellulose are no longer used because of regeneration problems, but are used with some success in the development stages of cephalosporin antibiotics [50]. Direct isolation by adsorption in a suitably designed treatment train can offer the advantages of high selectivity and yield. In general, non-polar macroporous resins are suited for the isolation of hydrophilic weak acids or bases as well as amphoteric or neutral molecules [50]. Neutral polymeric sorbents such as polyaromatics (Amberlite XAD-4, 16, 1180, Diaion HP20), aliphatic esters (Amberlite XAD-7) and nitrated aromatics (nitrated Amberlite XAD-16) proved

Table 3. Adsorbents for isolation of hydrophilic bio-products

Type	Nature	Typical example
Active carbon	Inorganic	Ambersorb (synthetic carbonaceous)
		Anthrasorb (from coal)
	Organic	Active carbon (from fossil fuel)
Molecular sieves	Inorganic	Zeolite
	Organic	Sephadex (G and LH), Biogel
Non-ionic	Inorganic	Silica gel, aluminum oxide, TiO
	Organic	Synthetic macroporous adsorbents, cellulose
Ion-exchange	Inorganic	Zeolites, ion-exchange crystals
	Organic	Strong acid or basic gel-type resin in salt form

effective for penicillins and cephalosporins, but the aromatic sorbents provide the highest sorption capacity for cephalosporin-C [51]. For strong acids and bases, ion-exchange chromatography may be more attractive. Strong acid and base exchanger and weak acid resin with single particle gel structure of styrene-divinyl-benzene (DVB) or methacrylate matrices and multiple particle aggregate (sponge like) structures of the macroporous type have been known for quite some time for hydorphilic cephalosporins [52]. Functionalized DVB-styrene-type resin has been exploited for purification of CPC [53]. Halogenated cross linked polymers of aromatic polyvinyl moiety, such as brominated Diaion HP20, can provide an effective method for recovery of cephalosporin-C [54]. Diaion and Amberlite resins adsorb and release ionic species based on hydro-phobic or hydrophilic interactions and they are usually employed under iso-cratic conditions. Since the medium contains various other anions, a single adsorbent can seldom be effective for a total purification protocol. In order to recover CPC in high yield and purity, adsorptive separation was practised using a series of resins in the sequence: Diaion WA-30 (weakly basic anion exchanger on polystyrene with alkylamine in OH form), HP-20, SK1B (strong acidic cation exchanger on polystyrene with sulfonic acid in gel form), Amberlite XAD-2000, and Diaion SK-1B [18]. Selectivity and capacity of a resin is greatly influenced by the nature of inorganic ions which also improve the separation selectivity in specific cases. Presence of alkaline earth cations augments separation of CPC from deacetyl CPC in neutral solution when macroporous non-polar resin is used [50]. Purification of 7-ACA produced by CPC acylase is possible based on preferential adsorption of the by-product, α-aminoadipic acid on a styrene-based resin (Diaion PA 408) that provides more than 95% purity of 7-ACA [55].

Lipophilic intermediates (see Sect. 3.2.1) of CPC can be purified with liquid ion-exchangers (Amberlite LA2-acetate) or solid resin (Amberlite IRA 68-acet-ate) [56] which essentially remove the bulk of the p-toluenesulfonic acid present as the unconverted reagent. It is also possible to use this resin for direct adsorption of CPC from a broth, even with a volume ratio between the resin and broth as high as 1/1. The lipophilic intermediates can be purified even at an aforesaid volume ratio as low as 1/10.

Recently, adsorptive separation of CPC has been demonstrated utilizing a column of Amberchrome polymeric reverse phase resin that exhibits highly favorable adsorption behavior [57] and a high degree of resolving power. Various reverse phase resins are commercially available [58], but their applica-tion in β-lactam separation is limited by the requirements of low pressure, chemical stability, column efficiency, and surface characteristics. Silica bonded with non-polar functional groups such as alkyl (butyl: C-4, Octyl: C-8, Oc-tadecyl: C–18) or aromatic phenyl is the common reverse phase support, but it is not very useful in cephalosporin separation because of their limited pH range (2–7.5) or application. In particular, solubility of a cephalosporin like 7-ACA is low and it is unstable on silica at alkaline pH. Porous graphite carbon made by impregnating a $7 \, \mu M$ spherical porous silica gel template with a melt of phenol and hexamine (6:1 w/w) has been demonstrated for separation of the diastereoisomer of the cephalosporin antibiotic, Anetil E 47 (Glaxo) and two pairs of isomeric impurities [59]. Highly cross-linked styrene DVB copolymer-type non-ionic macroporous resins (i.e., Amberchrome CG-161) are slowly replacing silica-based reverse phase packing by offering better chemical stability for gradient purification and recovery of biomolecules. The chemical properties of these resins are consistent throughout the bead and provide high selectivity of adsorption/desorption. Polystyrene-DVB gels can be effective [60] as these materials can eliminate the problems of pressure instability and incomparability with aqueous eluent associated with other columns and can withstand pressures up to $350 \, kg \, cm^2$.

3.2 Liquid-Liquid Extraction

3.2.1 Extraction via Lipophilic Intermediates

Though the application of organic solvent to whole broth extraction of penicillins and new antibiotics is successful [61], the microbiologically-derived cephalosporins, particularly those containing D-α-aminoadipamido side chain at C-7 position, are difficult to isolate by solvent extraction because of their amphoteric nature and high solubility in water. The naturally occurring CPC and the semisynthetic analogues like 7-ADCA, cephalexin, cephaloglycin, and cephalothin are amphoteric in nature. The attempt at a liquid-liquid extraction process is based on the chemical reaction of CPC leading to lipophilic derivatives which are extractable with solvent in the usual way. The objective is to suppress the basic character of the amino group in the side chain giving dibasic acids that are solvent-extractable at low pH. CPC derivatives reported in the literature [62] are N-isobutoxycarbonyl-, N-chloroacetyl-, N-benzoyl-, N-acetoacetyl-, N-dodecanyl-, N-isobornyloxy- carbonyl-, 2,4,-dichlorobenzyl-, butylcarbamoyl-, phenyl-carbomyl-, and 2,4,6-trinitrophenylcephalosporin C, and 3-acetoxymethyl-7 β [5-carboxy-5 (2,6-dimethyl-3,5-diethoxy-carbonyl-1,4-dihydropyridin-1-yl)] pentanamidocephalosporin-3-ene-4 carboxylic acid, which, after extraction from aqueous solution at low pH, can be easily transformed to 7-ACA. The preferred solvents, particularly for N-chloroacetyl and N-acetoacetyl, are ethyl acetate and n-butanol, respectively. The N-alkoxycarbonyl derivatives, i.e., N-isobornyloxycarbonyl obtained from alkoxyalkyl chloroformate are extracted by 4-methyl/pentane-2-one and n-butanol. It is possible to use n-butyl isocyanate and phenylisocyanate to form the corresponding ureas with CPC which can be extracted into n-butanol or ethyl acetate, respectively. The problem of such solvent extraction strategy, however, is the loss of solvents with aqueous phase because of their partial miscibility with water as a result of fairly high dielectric permittivity values of the solvents. Furthermore, production of the derivatives requires media which are rich in CPC or partially purified. For efficient extraction, it is necessary to combine the extraction with a preliminary chromatographic purification of the spent media.

In order to circumvent the above problems, lipophilic derivatives of CPC were obtained in good yield from alkyl and alkylsulfonyl chloride using unclarified media [56]. The formation of these derivatives namely N-tosyl- and N-p cumylsulphonyl CPC is independent of the concentration of CPC. These are sufficiently lipophilic to be extracted under acidic conditions (pH 2.0) with common organic solvents such as ethyl acetate and methylisobutyl ketone with purity ranging from 40 to 50%. The low purity is attributable to the reaction of other substances with sulfonyl chloride giving compounds containing the sulfonamide group. The derivatives can be purified by using liquid ion-exchanger (i.e., Amberlite LA2 acetate) or solid resin (IRA 68 acetate).

3.2.2 Extractive Esterification

Since spent media of CPC contain substantial amounts of deacetyl CPC, it is necessary to extract this side product in such a way as to maximise its use for further chemical modification. Processes involving low pH conditions for extraction will be unsuitable because deacetyl CPC easily forms a lactone which is not amenable for further synthetic modification due to the difficulty in opening the lactone ring. A process called "extractive esterification" can eliminate the above shortcomings and meet the requirements for an efficient extraction process [62, 63]. The process involves N-blocking the mixture of cephalosporin in an aqueous medium at slightly alkaline pH, followed by treatment with a solution of a diazoalkane in a water-immiscible organic solvent at acidic pH. A mixture of the biesters of N-blocked CPC and deacetyl CPC is formed and is extracted into the organic phase. In a typical system, the CPC mixture is acylated with ethylchloroformate at pH 8.0 followed by reaction with diphenyl-diazomethane in dichloromethane at pH 3.8 to give bis(diphenylmethane) ester of the cephalosporin in the organic phase. The relative volume of the phases can be adjusted so as to achieve considerable concentration of CPC. The anions such as phosphate and sulfate usually present in the media have low nucleophilicity and do not compete against carboxylate ions for attack on the protonated diazoalkane. Strong carboxylic acids such as cephalosporin acid are esterified preferentially in the presence of weak carboxylic acids. The dichloromethane solution containing the ester can be effectively utilized for subsequent chemical transformation to 7-ACA. Thus, both the products of CPC media can, in principle, be utilized for semisynthetic cephalosporins.

3.2.3 Reactive Extraction

The difficulty in solvent extraction of cephalosporin antibiotics is increased by their being ionic at high pH and unstable at low pH. In such cases, extraction accompanied by chemical reaction (with an extractant) called "reactive extraction" can provide an attractive method for separation of the antibiotics. Two mechanisms of reactive extraction of β-lactams can be suggested. In the first, known as ion-pair extraction, the extractant, typically a tertiary amine (A), dissolves in an organic phase and reacts with the β-lactam anions (P⁻) and a proton in the aqueous phase according to

$$A \text{ (org)} + P^- \text{(aq)} + H^+ \text{ (aq)} \rightleftharpoons A \, H \, P \text{ (org)} \tag{1}$$

The transport of the anion from one phase to the other requires the co-transport of a cation. The reaction is instantaneous and the extraction rate is controlled by the mass transfer of the ionic species.

The other mechanism, known as liquid-liquid ion exchange, involves water-insoluble extractant, and counter transport of a second anion occurs in order to maintain the electroneutrality. The removal of a β-lactam anion (P⁻) from the aqueous phase by an ion-exchange with the anion (Cl⁻) of the extractant QCl (typically a quaternary salt) dissolved in the organic phase takes place according to the reaction

$$QCl \text{ (org)} + P^- \text{ (aq)} \rightleftharpoons QP \text{ (org)} + Cl^- \text{ (aq)} \tag{2}$$

The extraction efficacy depends on the β-lactam type (dissociation constant), solvent, and the extractant through equilibrium relationship. The extractant should be able to provide stripping of the anion to another aqueous phase to effect separation. Table 4 gives a summary of reported studies on reactive extraction of β-lactam antibiotics. Under suitable physico-chemical conditions, reactive extraction can provide more than 90% recovery of penicillin-G directly from spent media [67]. Secondary and tertiary amines as the extractant for penicillin G/V provide adequate extraction stripping ability and no formation of a third phase. Other probable extractants are bis-(2-ethylhexyl) phosphoric acid and 2-hydroxy-5-nonylbenzophenone oxime (LIX 65N). CPC cannot be extracted with amine-type compounds because of difficulty in formation of ion-pair. It can be extracted from an alkaline into an organic phase containing Aliquat 336 and stripped into an acidic phase of acetate buffer without any appreciable decomposition [64]. Other amphoteric compounds like 7-ACA, 7-ADCA, and cephalexin are also amenable to reactive extraction with Aliquat 336 employed as 5% solution in butylacetate or kerosene [69].

3.3 Liquid Membrane Extraction

3.3.1 Emulsion Liquid Membrane

Since discovery two decades ago, the emulsion liquid membrane (ELM) has been extensively studied for the separation of metal ions, amino acids, and carboxylic acids from dilute aqueous solution [70]. ELM extraction does not encounter the equilibrium limitations of conventional solvent extraction

Table 4. Reactive extraction systems for β-lactam antibiotics

β-Lactam	Extractant/ carrier	Solvent	Mechanism	Reference
Cephalosporin-C	Aliquat-336[a]	Butyl acetate	Ion-exchange	
Penicillin-G	Amberlite LA-2[b]	Butyl acetate	Ion-pair extraction	[61] [65] [66] [67]
Phenyl and phenoxy acetic acid	Adogen[c]	Isoamyl acetate, Diiso-propyl ether, kerosene/n-hexane		
Penicillin V, Penicillin N	TBAS[d] Amberlite LA-2[b]	Hexane, octanol, dichloro-methane	-do-	[61] [68]
Clavulanic acid, amoxicillin	Aliquat-336[a]	Butyl-acetate	Ion-exchange	[61]
Olivanic acid	-do-	Butanol, hexane, dichloro-methane	-do-	[61]
7-Aminopenicillanic acid and 7-amino-cephalosporanic acid	-do-	Butyl-acetate	-do-	[69, 74]

[a] Tricaprylyl methyl ammonium chloride,
[b] Secondary amine
[c] Tri-octyl amine
[d] Tetrabutyl ammonium hydrogen sulfate

processes and can offer high selectivity of separation of solute from low concentration feed. Since natural cephalosporins exist in ionic form at considerably low concentrations, reactive extraction (Sect. 3.2.3) in a water/oil/water ELM via the so-called "facilitated transport" mechanism can serve as the basis for an effective separation method [71]. The separation is effected by simultaneous extraction-stripping processes occurring in a single step. Because of reactive extraction, high separation selectivity and transport flux can be achieved. However, toxicity of the carrier and solvent should be carefully considered while designing an ELM system for the separation of β-lactams.

There is practically no information on ELM extraction of cephalosporin antibiotics. However, the technique has been extensively studied for separation of penicillin-G from simulated dilute solution [72, 75]. Simultaneous separation and bioconversion of penicillin can be advantageously carried out in a system of enzyme-encapsulated ELM reactors [73] to produce 6-APA and ampicillin with high yield and purity. In principle, almost all natural cephalosporins and the semisynthetic analogues, i.e., 7-ADCA, cephalexin, etc., are amenable to reactive extraction in ELM [66]. Current studies at the author's laboratory revealed evidence of facilitated transport of CPC and 7-ACA in a suitably designed ELM system [74] which is now being studied for direct extraction of the medium. Since it contains structurally similar CPC and deacetyl CPC in appreciable proportions, selective separation is complicated from a practical point of view [62]. An ELM reactor proposed for production of 6-APA can perhaps be effective for 7-ACA from CPC using acylase enzyme now being developed [13]. A similar strategy can also be explored for enzymatic production of 7-ADCA where the substrate and by-product can be transported across the membrane via ion-pair formation with an amine-type carrier.

3.3.2 Supported Liquid Membrane

The liquid membrane involving microporous polymeric support (membrane), the so-called supported liquid membrane (SLM), can be more feasible for scale-up and adaptability for continuous operation. The liquid phase containing the extracting agent (in case of reactive extraction) is immobilized in the pores via capillary forces. SLM in a planar flat sheet configuration is more suited for fundamental studies of mass transfer mechanism as regards flux and selectivity measurements. However, the present easy availability of high surface area modular membrane units, i.e., hollow fiber (HF) membranes, opens up wide avenues for practical application of the SLM technique.

Reactive extraction in SLM has been studied for separation of penicillin-G from a model media [75] and a mixture with phenyl acetic acid [76] using tetrabutyl ammonium salt/tertiary amine and hydrophobic macroporous membranes as the carrier and support, respectively. A HF SLM system was utilized to study the combined extraction of penicillin-G and enzymatic hydrolysis to 6-APA [77] using tertiary amines as the extracting agent. However, membrane stability problems should be resolved before commercial application can be established. Separation of CPC has been studied at the author's laboratory [78] using an SLM system where CPC permeates from an alkaline feed of carbonate buffer into an acidic stripping solution of acetate buffer across a liquid phase of Aliquat-336 (extractant) and butylacetate (solvent) immobilized in a polypropylene support. However, the system suffers from the drawback of high mass transfer resistance across the liquid membrane as well as low membrane stability.

3.3.3 Non-Dispersive Extraction in Hollow Fiber Membrane

Non-dispersive solvent extraction utilises the immobilized liquid-liquid interface at the pore mouth of microporous membranes to effect phase-to-phase contact and the mass transfer process [79]. Extraction utilizing HF has been extensively studied for separation of pharmaceutical bioproducts [80]. Extraction and stripping are carried out usually in two HF membranes. A pH swing process based on reactive extraction with Amberlite LA-2 in two HF modules has been suggested for separation and concentration of phenoxyacetic acid [80], which is a by-product of semisynthetic cephalosporins. The combination of a hydrophobic membrane for extraction and a hydrophilic one for stripping exhibited higher separation efficacy and solute recovery. Such a strategy can be explored for recovery of cephalosporins from a filtered medium. Alternatively, non-dispersive solvent extraction of lipophilic intermediates of CPC (Sect. 3.2.1) and extractive esterification (Sect. 3.2.2) can be advantageously carried out in HF.

3.3.4 Emerging Liquid Membrane Processes

A hybrid operation of SLM, the so-called "Hollow Fibre Contained Liquid Membrane (HFCLM)", developed basically to eliminate the shortcomings of SLM, has been studied for extraction of carboxylic acid and other fermentation products [81]. The HFCLM technique has proved effective for separation of penicillin-G with more than 90% recovery via reactive extraction with tertiary amines [82]. The technique can perhaps be exploited to develop recovery processes for cephalosporins too.

A novel approach to avoid leakage and swellings of common ELM is to carry out the ELM extraction in a hollow fiber contactor. This approach, successfully demonstrated for metal ion removal from aqueous dilute solution [83], can, in principle, be effective for separation of cephalosporins from dilute media offering long term stability.

Rotating film pertraction (RFP) is another novel membrane technique in which the mass transfer occurs via "uphill transport" or transport against an apparent concentration gradient such that high product concentration can be achieved. This technique proved very effective for recovery of phenylalanine [84], and could be an option for cephalosporin antibiotics.

3.4 Membrane (Synthetic) Separation: UF/MF/RO/ED

Recent advances in the development of new membrane materials and module design/operation have resulted in fascinating opportunities for integrating membranes with various bioprocesses. This aspect has been discussed thoroughly in the literature [85]. The most widely studied application of the membrane in cephalosporin production is perhaps the ultrafiltration (UF), microfiltration (MF) and/or reverse osmosis (RO) for cell mass separation or concentration. A process has been described wherein cells are first removed by a 0.2 μM MF and then proteins and polysaccharides by a UF/diafiltration using a 10 KMWCO membrane [21]. The UF permeate is concentrated by RO and the antibiotic is finally purified by high performance liquid chromatography to achieve a recovery of 98.5%. For an industrial scale purification of CPC from the spent medium, integrated membrane separation has been proposed [18]. UF (Carbosep NMWL 50 000) for cell mass separation, column chromatography in ion exchange resins (WA-30, HP-20, XDA-200, SK-1B in sequence) for purification, and RO (300 MW) for product concentration have resulted in an overall CPC yield of 96% with 98% purity. UF and dialysis membranes have been frequently used at the preparative scale for concentration and desalting of enzyme extract during the biochemical investigation of natural and semi-synthetic cephalosporins [13,16,43,46]. MF membranes of polysulfone, PTFE, and cellulose acetate and UF membrane of polysulfone can provide high permeate flux and cell rejection from the spent medium of almost all cephalosporin-producing microorganisms [20].

Reverse osmosis has been commercially used for concentration of a mixture of 6-APA and phenylacetic acid in order to get the benefit of reduced solvent cost and increased efficacy of subsequent precipitation steps [7]. Electrodialysis with ion- exchange membrane is very effective for in situ removal of phenyl acetic acid (an inhibitory by-product) to obtain high yield of 6-APA during enzymatic hydrolysis of penicillin-G [86]. Such an approach can enhance the reaction rate by around 65% but at the cost of the overall yield because of loss of substrate by permeation across the membrane, which can, however, be compensated by increasing the penicillin-G conversion via the use of high concentration of the substrate in the feed. An identical approach appears feasible for the enzymatic production of 7-aminocephalosporanic acid and other semi-synthetic cephalosporins. Alternately, UF membrane bio-reactor studied for enzymatic synthesis of 6-APA [87] may also be examined for enzymatic synthesis of semi-synthetic cephalosporin. UF may be a cost-effective technique for recovery of almost all extracellular antibiotics, at least in the cross-flow mode of operation [88].

3.5 Aqueous Two-Phase Partitioning

Aqueous two-phase systems (ATPS) are formed by two different polymers, the most widely used being dextran and polyethylene glycol, or a polymer and a salt, in certain proportion. In an ATPS, a given molecule partitions between the two phases depending on its partition coefficient (K), defined as the ratio of the molecule concentration in the top phase to that in the bottom phase. The strong partitioning behavior of small biomolecules in ATPS may provide an efficient tool for their recovery from the spent medium [89,90]. The value of K for penicillin-G in polyethylene glycol (PEG) /phosphate two-phase system can be as high as 200 [91] which offer the possibility of ATPS finding practical application. Recovery of macrolide antibiotics such as Pristinamycin from a clarified medium of Streptomyces pristinaespiralis in a dextran/PEG ATPS has been reported recently [92]. Use of affinity ligand can enhance partitioning behaviour of vancomycin antibiotic in a similar ATPS [93].

The inherently difficult separation of structurally identical CPC and deacetyl CPC can be achieved in a PEG-6000/phosphate/ammonium sulfate two-phase system [19]. The experimentally observed K values (K > 1 for CPC, K < 1 for deacetyl CPC) may be the criteria based on which a process can be developed for this difficult separation. The interaction with the cells is found to be unimportant and viscous undiluted medium can be used immediately.

4 Method of Purification

4.1 Selective Precipitation and Crystallization

Generally, crystallization and/or precipitation is used in the final stages of concentration and purification of an antibiotic. Precipitation is preceded by extraction, the method of which may vary depending on the nature of the antibiotic. In case of penicillins, the usually practised solvent (butyl/amylacetate) extraction leads to a 120- to 150-fold enrichment of the product, which is finally precipitated from the extract as salt using organic bases like tertiary morpholine, N-ethylhexahydropicolins, and ethylpeperidine [94]. Direct precipitation of amphoteric tetracycline and oxytetracycline from spent medium using a long chain quarternary ammonium salt has also been reported [95].

Lipophilic intermediates of CPC, after extraction with suitable solvent and water washing, can be precipitated in the form of a sodium 2-ethyl-hexanoate and crystallized at a pH of 5.0. The semi-synthetic cephalosporins, i.e., cephradin, cephaloglycin and 7-ADCA, which are zwitterionic in nature in the pH range of 3.5–8, can be isolated as salt of hydrochloride, bromide, tosylate, nitrate, or hydrogen sulfate and crystallized from water at pH 1.0 by addition of isopropyl ether. The solubility of the semi-synthetic cephalosporins depends on the side chain and can vary from less than 0.1% to more than 30%, whereas their solubility in DMSO and DMF are greater than 20%. A convenient method for isolation of 7-ACA from the reaction mixture is to esterify to 7-ACA esters via phosphorus pentachloride-mediated reaction with benzoyl bromide, solvent extract and concentrate with methylisobutyl ketone, hydrolyse with alkali, concentrate in the aqueous phase, and finally crystallise at the isoelectric point. CPC can be similarly isolated as an N-chloroacetyl derivative with chloroacetyl chloride or chloro-acetic acid anhydride, and crystallized as quinoline salt, i.e., N-chloroacetyl CPC quinoline salt [50, 51]. The β-lactam in deacetoxy cephalosporin (i.e., 7-ADCA) is more stable than cephalosporanic acid and therefore their isolation is relatively simple. It is possible to form 7-aminodecetoxy-cephalosporanate ester from CPC by treatment with phosphorus pentachloride and dry pyridine. The 7-ADCA ester and cephalexin derivatives are separated via sulfuric acid treatment to precipitate the 7-ADCA ester from ethyl acetate extract. Acid hydrolysis of the estser then gives 7-ADCA. The CPC derivatives have limited solubility in water compared to CPC. Thus, separation of the derivatives by precipitation provides a general strategy for isolation of CPC in a practical situation.

4.2 Separation and Purification by Chromatography

The increasing application of chromatography in various forms to the analysis of β-lactam antibiotics indicates versatility of the techniques for large-scale processes [96]. In spite of being relatively costly, chromatographic methods continue to be dominantly practised on the preparative scale. The partitioning of β-lactams between the solvent and a solid can be derived from such phenomena as adsorption, molecular sizing (gel filtration), enzyme specificity (affinity chromatography), ion exchange, and hydrophobic interaction. Though the techniques are well developed as powerful analytical tools, the technology of large-scale chromatography (100–150 gm of loading capacity) has been developed for recovery of only a limited number of antibiotics [96,97].

4.2.1 Thin Layer and Paper Chromatography

Thin layer and paper chromatography (TLC and PLC), which offer great possibilities for physical separation, are applicable for analytical purpose only. A recent review [98] presents the TLC conditions for various cephalosporins like cefachlor, cefadroxyl, cephradine, cefamandole, cefoxitin, cefatoxine, cefaxitin, ceftizoxime, cephaloxin, and cephaloridine. The compounds were analysed using silica as the sorbent and various solvent systems like 0.1 mol/l citric acid – 0.1 mol/l potassium phosphate ninhydrin in acetone (60:40:1.5 v/v), acetic acid-water-butanol (1:1:4) and ammonium hydroxide methanol (1.5:100). In certain cases, i.e., cefachlor, sample pretreatment with 0.1 mol/l phosphate buffer is recommended because of solubility considerations. Cellulose on polyster film as the support provides excellent stability for cephalosporins [98] but silica gel is preferred when high loading is required. Thirty cephalosporins have been identified by using TLC on silanized silica gel layers containing a fluoresscence indiator [99]. Cellulose TLC is rapid and does not respond uniformly to sovent systems low in water content. Because of high polarity of CPC, its TLC with cellulose rather than silica is recommended. Both CPC and deacetyl CPC are separated using n-butanol-acetic acid-water (3:1:1 v/v) as the solvent system [48]. TLC of 7-ACA is somewhat difficult because of its relatively low solubility in water and stability on silica gel at high pH (pH > 7) at which 7-ACA is highly soluble. Automated cation exchange chromatography is the preferred choice for 7-ACA.

TLC on polyamide resin is useful for cephaloglycin and cephalexin. Because of the high hydrogen bonding nature of the adsorbent, improved separations result. Cellulose and silica can also be employed but only in aqueous acetonitrile solution. Cellulose is recommended for cephaloglycin derived from chemical sources but not for biological samples containing residual cephaloglycin. Resolution of less biologically active L-cephaloglycin from D-cephaloglycin is possible by TLC. Other uses of TLC reported are the detection of the tripeptide (ACV) and activity of deacetoxy CPC hydroxylase during growth [28,34].

Prior to successful implementation of high performance liquid chromatography (HPLC), reverse phase (RP) TLC of cephalosporin antibiotics was performed with chemically bonded dimethylsilyl silica gel in order to obtain a suitable HPLC separation system. The difference in solvent strength for efficient separation between TLC and HPLC can be attributed to the fact that, in HPLC, the solvent elution power acts in an isocratic manner whereas, in TLC, it acts in a gradient manner. RPTLC of 7-ACA, cephalothin, cephalexin, cephaloglycin, and cephaloridine were performed in methanol-water solvent-silica gel 60 HF_{254} (silanized) system [48, 100]. Ion-pair chromatography has been applied to RPTLC of these cephalosporins. Deacetoxy CPC and deacetyl CPC formed in media could be separated by RPTLC carried out using opti-UpC_{12} (Fluka) and 25 mmol/l NH_4HCO_3 as the developing solvent [40]. A correlation of mobility between TLC and HPLC separation obtained for cephalosporins indicates a possibility of direct transfer of chromatographic system from TLC to HPLC.

Paper chromatography (PC) is easily adopted for separation of water soluble antibiotics, but not for strongly lipophilic substances. PC can allow greater migration distance and thus achieved better separation. Because of such advantages, PC was preferred over TLC before the other HPLC techniques and plastic films as support materials for TLC were developed. Data on PC of cephalosporins are available in the literature [101]. The important examples are separation of CPC from related substances like penicillin N and/or 6-aminopenicillanic acid (system: *n*-butanol-acetic acid-water, 4 : 1 : 5 on Whatman paper No. 1), deacetyl CPC, and CPC lactones, and identification of semisynthetic analogues like cephalothin, cephaloglycin, and cephaloridine in biological fluid.

4.2.2 Ion-Exchange Chromatography

Ion-exchange chromatography, generally performed with either a continuous or step-wise elution gradient, involves interaction of charged functional groups with ionic functional group of opposite charge on the adsorbent surface. Ion-exchangers are derived from substrates like dextran (Sephadex), agarose (Sepharose), cellulose (Sephacel), and polystyrene-DVB (Amberlite, Diaian). Weak anion-exchanger is provided by the diethylaminoethyl (DEAE) group whereas the quaternary aminoethyl (QAE) group is a strong anion exchanger. Weak and strong cation exchangers are based on carboxy methyl (CM) and sulfoxypropyl (SP) groups, respectively. Trisacryl plus ion exchanger consisting of the matrix poly (*N*-tris[hydroxymethyl] methylmethacrylamide) provides stability over a wide pH (1–14) range. Very limited information is available on recovery of antibiotic products using ion-exchange chromatography [70]. Macroporous weak cation exchanger with a carboxylic acid functional group can be used for purification of various antibiotics. Weak anion exchanger with a polyamine functional group (i.e., IRA 35-macroreticular, IRA-68 gel form) have been used for isolation of CPC over a pH range of 1–9. Quaternary ammonium-type ion-exchangers, i.e., Lewatit MP 500 A, has been suggested for separation of antibiotic-related compounds, i.e., phenyl acetic and phenoxyacetic acids [62]. Diastereomeric zwitterionic antibiotics have been separated on an anion-exchange column of resin in Cl^- form with an aqueous pH gradient (pH 7-1) generated using HCl [102]. Isocratic separation was done with 0.1 N acetic acid with a resin in acetate form. The separated products were found to retain the β-lactamase inhibitory and antibiotic activities. Cephalosporin derivatives can be separated by dynamic anion-exchange chromatography using *heteroions* like tetraethylbutyl ammonium, tetrabutylammonium, tetrapentylammonium and tetraoctylammonium ions [103]. Use of polystyrene resin containing the copper complex of lysins derivative has been reported for separation of CPC [104]. During biochemical investigation of cephalosporins, ion-exchange chromatography is almost invariably used. Table 5 lists typical examples which mostly comprise desalting (ion-exchanger elute) and purification of enzymes involved in the biosynthesis.

4.2.3 Hydrophobic Interaction and Hydroxyapatite Chromatography

Hydrophobic chromatography is the resolution and purification of water-soluble as well as lipophilic biomolecules on the basis of the hydrophobic nature of their surfaces. A variety of resins of alkyl and aminoalkyl agarose, useful for hydrophobic chromatography, can be made via different bonding techniques. It is generally possible to increase the contribution of hydrophobic interaction by increasing the number of carbon atoms in the hydrocarbon chain. Hydrophobic chromatography as a variation of the reverse phase technique has been used for the CPC derivatives using a large particle (37–50 μM) medium efficiency C_{18} column [105]. By using a small particle (10 μM) column prepared by chemically bonding octadecyltrichlorosilane to silica by an in situ technique [105], the separation efficacy for deacetyl CPC and deacetoxy CPC could be improved. However, all available columns are not satisfactory because of poor efficiency and the broad asymetrical resolution (peak) obtained. In situ bonding of mono-, di- and tri-functional octadecyl agents onto 10 μM silica of 60 A and 100 A pore diameters provide an improved method when a mobile phase of water containing potassium dihydrogen phosphate as an ionic modifier is used [106]. The column can be routinely used for the analysis of CPC medium. Cephalosporin synthesizing enzyme purified by hydrophobic chromatography on phenyl Sepharose CL-4 is reported to have good yield and activity [41, 44].

Hydroxyapatite chromatography has been used as the final step in the chromatographic purification protocol for deacetoxycephalosporin synthase [40,41], α-aminoacid ester hydrolase, and glutaryl 7-ACA acylase [13] using commercially available columns (i.e., Bio-Rad HTP/CEX-Sepharose 4B) and a stepwise elution technique. The enzyme elution is independent of molecular

Table 5. Ion exchange chromatography of cephalosporins and related enzymes

Compounds and source 1	Application example 2	Column 3	Mobile phase 4	Reference 5
δ-(L-α-Aminoadipy)-L-cysteinyl-D-valine (ACV) synthetase (ACVS) from *S. clavuligerus*	ACVS purification and desalting from crude extract	Mono Q HR 5/5 amino exchange	Elute: linear gradient of 0.1 to 0.5 mol/l MOPS buffer (morpholine-propane sulfonic acid/KOH) pH 7.5	[28, 29]
ACVS from *C. acremonium*	ACV isolation and desalting	Sephadex G-25 Dowex 50 X (H⁺)	100 mmol/l (MOPS) Trichloroacetic acid-NaOH, pH 4.5	[40]
ACVS from *A. nidulans*	ACVS purification	Sephadex G-150 Mono Q HR 10/10 FPLC DEAE sepharose	1 mmol/l pyridine, imidazole/HCl, pH 7.2 Tris-HCl, pH 7.5, 1 mmol/l dithiothreitol (DTT), 0.1 mmol/l EDTA, 10% glycerol Elute: 50–250 mmol/l NaCl gradient Above buffer Elute: 0–200 mmol/l, NaCl gradient	[30]
Isopenicillin-N synthase (IPNS) from *C. acremonium*	Desalting and IPNS purification from crude extract	DEAE-5 PM ion exchange HPLC Phenyl Sepharose CL-4B DEAE-Sepharose CL-6B Sepharcryl S-200	50 mmol/l tris-HCl, 10 mmol/l MgSO₄, 10 mmol/l KCl pH 7.4 0.01 mmol/l MOPS/NaOH,	[31]
IPNS from *S. clavuligerus*	Desalting and IPNS purification	Sephadex G-25 DEAE-5 PM (HPLC ion exchange) DEAE-Trisacryl.	250 mmol/l tris-HCl, pH 7.0 10 mmol/l MgSO₄, 10 mmol/l KCl 10% sucrose in tris-HCl, pH 7.0 0.05 mol/l tris-HCl, pH 7.2 0.1 mol/l DTT, 0.01 MEDTA	[34, 35]
IPNS from *A. chrysogenum*	Purification	Mono Q 5/5 (ion exchange FPLC)	220 mmol/l NaCl in 20 mmol/l tris-HCl, pH 7.2	[33]
IPNS (gene) from *A. nidulans* and *S. lipmani*	Desalting and IPNS purification from crude extract	Sephadex G-25 Cellofine A-800 Q-Sepharose	25 mmol/l MOPS, pH 8.0 25 mmol/l MOPS, pH 8.0 NaCl linear gradient	[36]

A.C. Ghosh, R.K. Mathur and N.N. Dutta

Table 5. (Continued)

Compounds and source 1	Application example 2	Column 3	Mobile phase 4	Reference 5
Isopenicillin-N epimerase (IPNE) from *S. Clavuligerus*	-do-	DEAE-cellulose (Whatman DE-52) Sephadex G-200	10 mmol/l pyrophosphate HCl, pH 8.0, 0.1 mmol/l DTT, 0.15 mol/l NaCl	[38]
		Mono QHR 75/5	10 mmol/l pyrophosphate, pH 6.0, 0.01 mmol/l DTT, 10 mol/l pyridoxal 5-phosphate 10 mmol/l pyrophosphate HCl pH 8.0 + as above	
IPNE from *N. lactamdurans*	-do-	Mono QHR 5/5 FPLC	20 mmol/l tris-HCl, pH 8.0, 0.05 pyridoxal phosphate, 0.1 mmol/l DTT	[39]
		Superose 12	50 mmol-tris-HCl, pH 7.0 1 mmol/l phenylmethyl sulfonyl fluoride (PMSF), 1 mmol/l DTT, 0.01 mmol/l pyridoxal phosphate	
Deacetoxy cepha losporin-C synthase/hydroxylase	Desalting and purification of DAOCS from crude extract	DEAE-Trisacryl DEAE-celluose/sepharose (Saphacryl S-200)	50 mmol/l tris-HCl, pH 7.0, 10% glycerol, 10% ethanol, 1 mmol/l ascorbate	[41]
(DAOCS)/(DAOCH) from *C. acremonium*	(recombinant enzyme also)	Mono Q 5/5 (FPLC)		
DAOCS from *S. clavuligerus*	Desalting and purification	Sephadex G-25	0.125 mol/l tris-HCl, pH 7.2 in 0.05 DTT, 0.01 mmol/l disodium EDTA	[43]
		DEAE-trisacryl	-do-	
		Mono Q HR 5/5	0.05 mol/l above buffer, 0.07 mol/l KCl, 0.25 mol/l sucrose	

Enzyme/source	Purpose	Column	Buffer/conditions	Ref.
DAOCS/DAOCH from *S. clavuligerus*	Purification and separation of DAOCS/DAOCH activity	DEAE-trisacryl	0.05 mol/l Tris-HCl, pH 7.2 1 mmol/l DTT, 0.01 mol/l EDTA	[34]
DAOCH from *S. clavuligerus*	Desalting and purification	Sephadex-75	20 mmol/l tris-HCl, pH 8.0, 2 mmol/l DTT: 0-2 mmol/l KCl linear gradient	[44]
		Mono Q 16/10 (strong anion exchange FPLC)	above buffer: 80–180 mol/l NaCl linear gradient	
DAOCS from *S. lactamdurans*	-do-	*Sephadex G-75*	20 mmol/l tris-HCl, pH 7.0 0.1 mol/l DTT, 1 mol/l PMSF	[45]
		DEAE-Sephacel	300 mmol/l tris-HCl, pH 9.5: pH 9.5–4 linear gradient	
α-Amino acid ester hydrolase from *Acetobacter turbidan*	Purification	CM-Cellulose (cation exchange)	0.1 mol/l Phosphate, pH 6.2: KCl gradient in buffer	[8]
		DEAE-Cellulose (Whatman DE-51) (anion exchanger)		
Glutaryl-7ACA acylase from *Pseudomonas sp.*	-do-	Q-Sepharose	20 mmol/l tris-HCl, pH 9.0: 0–0.6 mol/l NaCl linear gradient	[13]
		DEAE-Trisacryl	20 mmol/l Maleate, pH 5.8: 0–1 mol/l NaCl gradient	
		Toyopearl HW65F/55F	100 mmol/l phosphate, pH 7.5 $(NH_4)_2SO_4$: 35–0% linear gradient	
Glutaryl-7ACA acylase and Cephalosporin-C acylase from recombinant *E. coli* of *Pseudo sp.*	-do-	CM-Toyopearl 650M (anion exchanger)	Buffer as above: 0–0.2 mmol/l NaCl linear gradient	[10]
		DEAE-Toyopearl	200 mmol/l tris-HCl, pH 7.2: 0–0.2 mol/l NaCl linear gradient	
Cephalosporin-C deacetylase from *Bacillus subtillis*	-do-	DEAE-Toyopearl 650M	tris-HCl-0.5 mol/l NaCl: 0.05–0.015 mol/l NaCl linear gradient	[16]

weight, thus making it a strong tool for final separation and purification of the enzyme.

4.2.4 *Gel Filtration*/Size Exclusion Chromatography

Separation by gel filtration depends on the different abilities of sample molecules to enter pores of the packing materials. The traditional gel filtration media, i.e., cross-linked dextran gel and the cross linked polyacrylamide gel are now available in a wide range of porosities. Application of high performance techniques to gel filtration has enhanced its resolving power. Table 6 lists some examples of the use of gel filtration chromatography for purification and molecular mass determination of enzymes involved in cephalosporin biosynthesis. In most of the cases, the technique is employed as a final purification step to obtain enzymes with high specific activity [8, 13, 30]. Other examples of the use of gel filtration chromatography are the analysis of semisynthetic cephalosporins, i.e., ceftazidime and cepholothin in commercial samples [107].

4.2.5 *High Performance Liquid Chromatography*

High performance liquid chromatograpy (HPLC) offers the advantages of superior resolution and speed of operation. Nearly all new antibiotics are studied from development to clinical application by using HPLC. There are several techniques for the analysis of cephalosporin by HPLC as discussed in a recent review article [98]. Various types of column materials are available, but only a few have been used for analysis of cephalosporins and derivatives such as cephalexin, cefazolin, cephradine, cephaloglycin, cephalothin, cephaloridine, cefatrizine, cefactor, cefadroxil, cefoxitin, cefotaxime, cefamandole, cefuroxime, cefoperazone, and cefmetazole. These are based on silica octadecylsilane chemically bonded to silica and strong cation-exchanger bonded to silica and are commercially available as μ-Bondapak C_{18}, Zorbex C_8, Lichrosorb RP-8, Supelcocil LC-18, and Chromosorb LC-8. The UV detection wavelength of 254 nm is common for almost all the derivatives except for Cefuroxime (274 nm).

A method based on precolumn derivatisation and simple elution gradient has been developed for convenient determination of various compounds involved in biosynthesis of cephalosporin [108]. This involves derivatisation with a chiral flurogen (O-phthaldialdehyde with an optically active thiol N-acetyl-L-cysteine) followed by HPLC on a reverse phase column of Hypersil ODS. It is possible to determine simultaneously most of the common L-aminoacids, i.e. L- and D-α-aminoadipic acid, L-and D-valine, δ-(L-α-aminoadipyl)-L-cysteine, δ-(L-α-aminoadipyl)-L-cysteinyl-D-valine (ACV), glutathione, isopenicillin N, penicillin N, desacetoxy CPC, deacetyl CPC, and CPC. In a modification of the method, the sulfhydryl group of the tripeptide (ACV) was converted to the fluorescent derivative of monobromamine which was separated on a μ-Bondapak C-18 HPLC column and monitored with a fluorescence detector for estimation of ACV [33]. Some other examples are given in Table 7.

Reverse phase HPLC has been demonstrated to be more effective for cefazolin, cephalothin, cefoxitin, cefotaxime, cefamandole, cefuroxime, and cefoperazone in biological fluids [109]. A μ-Bondapak C_{18} column and a mobile phase of 0.01mol/l acetate buffer (in methanol/acetonitrile) was used with samples prepared by extraction and reextraction into an aqueous phase. Closely related methods for determination of cephalosporin derivatives in plasma and urine samples have been described in either reverse phase or the ion-pair mode using Lichrosorb RP-18 and methanol-phosphate as the column and mobile phase, respectively. Ion-pair RP HPLC method is more effective for sharp separation of cephalothin from 7-ACA in a Lichrosorb RP-2 column with methanol-water (1:4) as the mobile phase [116]. 1-Heptane sulfonic acid (pH 3.5 with acetic acid) and tetrabutylammonium phosphate (pH 7.5) were used as the ion-pairing agents. Ion-pairing of μ-Bondapak C_{18} with 0.6% cetyltrimethyl ammonium bromide in a borate-n-PrOH (10:3) solution (pH 8.5) has been described for eight cephalosporins [110]. Separation of various components of CPC media, i.e., CPC, desacetyl CPC, deacetoxy CPC, and thiomethyl CPC has been achieved using a similar technique [111]. The method can be easily automated and used for simultaneous product potency and by-product analysis. Ceftizoxime in biological fluid can be determined by ion-pair RP HPLC in an octadecyl column with K_2PO_4 buffer and tetrabutylammonium dihydrogen phosphate [112].

A reverse phase microparticulate C_{18} column and a mobile phase of water have been tested for preparative scale separation of cenfonicid disodium salt [7-D-α mandelamido-3-(1-sulfomethyl tetrazole-5-yl-thiomethyl)-3-cephem-4 carboxylic acid] from impurities (tetrazole sulfonic acid sodium salt) not readily removal by liquid-liquid extraction [97].

4.3 Electrophoretic Techniques

The utility of electrophoresis in separation of cells and biological molecules has been discussed in the literature [98]. Electrophoresis is capable of resolving molecules on the basis of their differences in molecular weights, electrophoretic mobilities, isoelectric points (PI), or various combinations of their properties. Analytical electrophoresis is done on a support matrix of paper, cellulose acetate, starch, agarose, or polyacrylamide gel. Agarose and polyacrylamide gel provide a porous matrix and they separate by charge and molecular weight. Zonal electrophoretic methods utilize a buffer with a pH above the PI of the ampholyte in the sample and separation is based solely on net charge. There is scope for application of large-scale zone electrophoresis for downstream processing of spent media utilizing continuous annular rotating flow, thick film continuous flow, and recycle continuous flow apparatus [113].

A.C. Ghosh, R.K. Mathur and N.N. Dutta

Table 6. Gel filtration chromatography of cephalosporin synthesising enzyme

Compounds and source 1	Application example 2	Column 3	Mobile phase 4	Reference 5
ACVS from S. clavuligerus	Desalting of crude extract	Sephadex G-10	0.1 mol/l MOPS, pH 7.5	[29]
ACVS from A. nidulans	Enzyme purification	Sephadex G-25 Ultrogel ACA-34	0.05 mol/l KCl, 20% w/v glycerol 50 mmol/l tris-HCl, 1 mmol/l DTT, 0.1 mmol/l EDTA, 9% w/v glycerol	[30]
	Final purification for molecular mass determination	DEAE Fast Flow (native as well as FPLC)	–do–	[30]
IPNS from C. acremonium	Purification and molecular weight determination	LKB ultrogel AC A54	50 mmol/l tris-HCl, pH 7.2, 0.1 mmol/l MgSO₄, 10 mmol/l KCl, pH 7.4	[31]
	Partial purification of fractionated enzyme	Sephacryl S-200	50 mmol/l tris-HCl, pH 8.0	[32]
IPNS from S. clavuligerus	Purification and molecular mass determination	Superose 12HR 10/30	0.05 tris-HCl, pH 7.2, 0.1 mmol/l DTT, 0.01 mmol/l EDTA	[34]
IPNS from S. lactamdurans	Final purification	Sephadex G-75	5 mmol/l MgSO₄, 5 mmol/l KCl, 5% w/v glucose	[35]
IPNE from S. clavuligerus	Purification and molecular weight determination of partially purified enzyme	DEAE-Affi-Gel	0.1 mmol/l OTT, 0.1 mol/l NaCl 0.1–0.3 mol/l NaCl linear gradient	[38]
		Sephadex G-200	10 mmol/l potassium phosphate, pH6.0, 0.1 mol/l DTT, 10 mol/l pyridoxalphosphate	
IPNE from N. lactamdurans	Purification and	Sephadex G-75	50 mmol/l tris-HCl, pH 7.0,	[39]

Enzyme/source	Purpose	Column/gel	Buffer	Ref.
DAOCS/H from C. acremonium	molecular weight determination	Superose 12HR 10/30 FPLC	1 mmol/l PMSF, 1 mmol/l DTT, 0.05 mmol/l pyridoxalphosphate phosphate	[41]
	Co-purification and molecular weight determination	Sephacryl S-200; Bio-Gel-A	50 mmol/l tris-HCl, pH 7.2, 10% glycerol, 10% ethanol, 10 mmol/l ascorbate	[40]
DAOCS from C. acremonium	Purification	Superose 12	20 mmol/l tris-HCl, pH 8.0 10% glycerol, 10% ethanol, 10 mmol/l ascorbate	[40]
DAOCS from S. clavuligerus	Purification and molecular mass determination	Superose 12HR 16/52 FPLC gel	0.05 mol/l Disodium EDTA	[43]
DAOCS/H from S. clavuligerus	-do-	Sephadex G-200 (Superfine gel)	0.05 mol/l tris-HCl, pH 7.0 1 mmol/l DTT, 0.01 mmol/l EDTA	[42]
DAOCS from S. lactamdurans	-do-	Sephadex G-75	20 mmol/l tris-HCl, pH 7.0, 0.1 mmol/l DTT, 1 mmol/l PMSF	[45]
Cephalosporin-C acylase from Pseudomonas sp.;	Purification	TSK-gel DEAE 5PW	20 mmol/l tris-HCl, pH 8.0	[14]
Glutaryl-7ACA acylase from Pseudomonas sp.	-do-	Sephacryl 300 (superfine gel)	150 mmol/l tris-HCl, pH-8.0	[13]
-do-	-do-	TSK gel G 2000SW	0.1 mol/l phosphate, pH-8.0 45% saturated (NH₄)₂SO₄	[10]
Cephalosporin-C deacetylase from B. subtillis	Purification and molecular weight determination	DEAE Sephadex A-50; Sephacryl S-300; Sephacryl G-200	50 mmol/l tris-HCl, pH 8.0 0.1 mol/l NaCl	[16]
α-Aminoacid ester hydrolase from Acetobacter turbidans	Final purification	Sephadex G-200	0.1 mol/l potassium phosphate, 20% (NH₄)₂SO₄, pH 6.2	[8]

Table 7. HPLC analysis during cephalosporin biosynthesis

Enzyme and source 1	Application example 2	Column 3	Mobile phase 4	Reference 5
Isopenicillin-N synthetase from *A. chrysogenum*	Enzyme assay	Reverse phase Whatman Partisil ODS-3 C_{18}	0.1% trifluoroacetic acid, 4–76% acetonitrile	[33]
Deacetoxy cephalosporin-C synthetase/hydroxylase from *S. clavuligerus*	Assay of enzyme activity and product	μ-Bondapak C_{18}	25 mmol/l NH_4HCO_3, 50 mmol/l sodium phosphate	[44]
Deacetoxy cephalosporin-C synthetase from *S. lactamdurans*	Separation and analysis of deacetoxy and deacetyl cephalosporin-C and enzyme assay	μ-Bondapak	10 mmol/l acetate, acentonitrile (99% v/v)	[45]
		μ-Bondapak NH_2	2% acetate, 4% methanol, 6% acetonitrile	[41]
Cephalosporin-C acylase from *Pseudomonan sp.*	Enzyme assay	Intersil ODS-2	0.56 g/l Na_2HSO_4, 0.36 g/l $K_2H_2SO_4$, 2–4% Methanol	[14]
	–do–	Lichrospher 100 RP-18	0.15mol/l citric acid, 5 mmol/l 1 sodium hexanesulfonate, 14% acetonitrile	[14]
	Peptide mapping	Reverse phase 214TP 5215	0–60% acetonitrile	
Glutaryl-7ACA acylase/cephalosporin-C acylase from *Pseudomonas sp.*	7-ACA formation	Intersil OPS-2	0.56 g/l $Na_2H_2SO_4$, 0.36 g/lKH_2PO_4, 2–4% methanol	[10]
Glutaryl 7-ACA acylase *Pseudomonas sp.*	Enzyme assay	Reverses phase $5MC_{18}$	10 mmol/l octylsulfonic acid, 10% methanol	[13]

4.3.1 Polyacrylamide Gel Electrophoresis

The most widely studied methods are polyacryl amide gel electrophoresis (PAGE) and gel electrophoresis in the presence of sodium dodecysulfate (SDS-PAGE). Various systems for PAGE are available and a summary of the applications of electrophoresis in the field of antibiotics appears in the literature [115]. Basic, acidic, and amphoteric antibiotics in general can be analysed with the main application in separation, identification, preparation, and purity control. Bioautographies for paper, agar gel, and gel electrophoresis have been described for detection of antibiotics using various test organisms such as *Bacillus subtillis*, *Staphylococus aureus*, *Escherichia coli* or *Micrococus flavus*. Preparative paper electrophoresis using 10% acetic acid as buffer afforded pure 7-α-methoxy CPC and detection of cephalosporins in spent media [115]. Compounds like CPC and deacetoxy CPC were detected by exposure to UV light, by coloration with ninhydrin reagent, and by the halo formed on the plate seeded with *Comamonas ferrigenaa*. Electrophoresis on Whatman paper No. 10 (pH 1.8, 2.5 h at 4 KV) was performed for ACV via radioactivity determination of the ninhydrin-stained peptide [40].

Analytical electrophoresis has been applied to the separation and molecular mass determination of various cephalosporin synthesizing enzymes mentioned in Table 5. It may be noted that preparative electrophoresis in vertical tube (12 cm length, 2–4 cm internal diameter) applied to purification of IPNS enzyme [32] can raise the homogeneity 1.5-fold but at a loss of 60% activity. Native gel electrophoresis has also been used as a final step in the purification protocol for glutaryl-7-ACA acylase [13] from *Pseudomonas sp.* IPNS genes from recombinant *E coli* have been purified and recovered by electroelution for direct sequence analysis [36].

4.3.2 Isoelectric Focussing

Isoelectric focussing has mostly been utilized to determine the isoelectric points (PI) of various cephalosporin-synthesizing enzymes using various protein standards. In a typical application, the acylase enzyme preparation was applied to a thin layer of 4% polyacrylamide gel containing 20% ampholine (pH 3.5–10) and electrofocussed for 2 h at 100 V after staining the gel with Commassie Blue R-250 [13] or silver [39]. Two-dimensional electrophoresis has also been reported for the study of IPNE from *N. lactamdurans* [39], but is of limited general applicability due to simultaneous detection of two PIs for the enzyme of same molecular weight. Thus use of a two-dimensional electrophoresis for preparative scale separation appears to be prohibitive.

5 Process Design and Scale Up Consideration

The methods discussed above have been tested mostly on the laboratory scale. The success of any separation/purification method depends on a clear understanding of the scale-up and design principle and availability of custom-made equipment. Very often improved versions/designs of equipment have to be sought in order to derive the benefit of reduced operating cost. Scale-up is not a simple multiplicative procedure because many process variables change while translating a laboratory scale process to a pilot plant.

Owing to the complex nature of a cephalosporin medium, its processing to recover/isolate the product requires a combination of methods. Furthermore, the microorganism must remain active and viable. Optimisation of the overall yield of isolation is essential because even a 1% increase in yield of CPC for a plant of 100 T/annum at a product cost typically of $100/kg will result in an extra annual benefit of $1 00 000 which is quite attractive. Scale-up and process design studies are prerequisites for such an optimisation which is, however, a costly venture.

Several chromatographic purification processes are amenable to easy scale-up and design. The technology of large-scale chromatography (100–150 gm of loading capacity) has already been developed [97] particularly using reverse phase materials. Since the preparative chromatography is performed at much higher sample column ratios than those used in analytical columns, it is important to select a chromatographic system under load conditions similar to those used on the process scale rather than those used in the analytical system. The column load (g sample/g HPLC support) primarily determines the process throughout and product purity. By properly optimizing the column load in laboratory experiments, the results can easily be translated to large-scale processes [97]. The commercialisation of macroporous resin adsorbents opens up wide possibilities for translating analytical procedures into pilot plant demonstration through a knowledge of the elution and regeneration properties, selectivities, and hydraulic behavior of the adsorbents. Perfusion chromatography, where convective flow dominates, can give faster separation and offers high potential for scale-up [96].

Liquid-liquid extraction in general is a mature technology and at least 25 different types of extraction equipment are in commercial use [116]. While the use of an extractor in biotechnology is limited to only a few types of [61], the selection and scale-up for extraction of cephalosporin antibiotics has not yet been dealt with. The ideal choice of equipment for extraction of cephalosporins via lipophilic intermediates, "extractive esterification", or reactive extraction is perhaps the Karr reciprocating plate column or the Kuhni-type column with a radial flow turbine-type impeller. These extractors have high hydraulic capacity and volumetric efficiency, defined as the ratio of the total specific throughout to height equivalent of one theoretical stage (HETS), and can be scaled-up from small test sizes to commercial columns in a simple, reliable manner. Indeed, the performance of Karr and Kuhni columns for continuous reactive extraction and reextraction of β-lactam could be simulated by a simple model consisting of a cascade of ideal mixing cells and extraction kinetics for which the model parameters, like number of stages and mean residence time, could be determined by separate measurement. Such a model and identified parameters could be used for scale-up and design calculations. However, it may be necessary to evaluate the effect of various constituents of a spent medium on the design and model parameters of the extractor. In fact, the column design is based on pilot scale trial with spent media that would provide HETS values as a function of the operating conditions for a particular size of column.

Liquid emulsion membrane (LEM) can perhaps function best for natural cephalosporins which are usually present at considerably low concentration in the spent media. However, it is not clear where LEM should belong in the downstream processing train. If LEM is to be used early in downstream processing for crude separation and concentration, it must be demonstrated that LEM operation can compete economically with well-proven commercial processes like adsorption chromatography. If LEM is to be used later in the downstream processing train, it must be shown that adequate and specific separation can be achieved and product loss would be acceptable. An economic analysis based on reliable pilot plant data should be carried out if LEM process is to gain acceptance as a commercially viable downstream operation. The major steps in an ELM process are the emulsion formation (emulsification), emulsion splitting, and ELM extraction. Though the selection and design of the equipment for the first two steps add decisively to the feasibility of the whole process, selection of suitable equipment for the third step is more crucial as the separation by ELM is based on a rate process rather than on an equilibrium relationship. In view of the high cost of the cephalosporins, a high degree of separation is essential. Thus for large-scale application, a counter-current column operation may be preferable to a mixer-settler operation. Furthermore, since renewal of the surface of the emulsion drop has significant influence on the permeation kinetics, the Kuhni-type of column which can provide repeated coalescence and redispersion of the emulsion drops should be suitable for the ELM process. This device is amenable to easy scale-up and design. However, it may be necessary to evaluate the effect of the various constituents of a spent medium on the design parameters of the extractor.

For non-dispersive reactive extraction in a hollow fiber contactor, mass transfer correlations were found to be valid for a variety of modules, suggesting easy scale-up [79], but introduced novel optimisation problems in terms of modular assembly configuration. Expected low height of transfer unit (HTU) value of HF device would suggest new perspectives in equipment useful for extraction.

As far as membrane (RO/UF/MF/ED) separation is concerned, there is no tailor-made membrane for a specific job. The design variables such as module hydrodynamics, flow rate, feed characteristics (as regards solute concentration, diffusivity, osmotic pressure, and viscosity) operating pressure, and fouling potential as function of product recovery can seldom be set a priori. Module performance for a complex media with highly fouling potential needs to be evaluated in a pilot plant to obtain design information for a full-scale unit and economic data for an integrated scheme. The flux decline/fouling attributable to particulate and bacterial adhesion must be quantified from long-term operation. The cost of membrane processing may vary depending on the type of module and the process integration adopted.

Industrial scale aqueous two-phase partitioning in multi-stage operation (for maximizing the yield) can rely on a liquid-liquid partitioning column [11]. However, the system-dependent hydrodynamic and mass transfer parameters should be evaluated experimentally to arrive at reliable scale-up and design criteria. Other issues such as phase separation, recovery of phase components (i.e., PEG/dextan) and salt for recycle, and recovery of cell mass without loss of activity should be addressed at the scale-up stage itself. This becomes an optimisation problem in as much as the degree of phase separation, which alone may affect the overall yield, indeed so observed for cephalosporin extraction [19], can be considered critical for the process. In order to assess the process reliability, use of on-line monitoring systems during scale-up studies has been emphasised to investigate not only the system parameters but also the product parameters such as activity and concentration [118]. Recently developed modelling techniques [119] are expected to provide much insight into the partitioning process and facilitate prediction of the separation at much lower scale-up efforts.

Though some efforts have recently been made to develop an industrial scale electrophoretic apparatus for downstream processing in biotechnology, large-scale application of the technique is yet to be established. The process is complex, involving ionic migration, diffusion, convection, and dissociation reactions which couple each other. The basic criteria and data for design are not adequately known. Efforts made modelling and simulation of continuous flow electrophoresis of protein [114], however, indicate the incentive for exploiting the technique for further investigation.

Acknowledgement. Financial support from the Department of Science and Technology, New Delhi vide sanction No. III-4(15) 94-ET has been gratefully acknowledged.

6 References

1. Newton GGF, Abraham EP (1955) Nature 24: 443
2. Martin JF, Liras P (1989) Enzymes involved in penicillin, cephalosporin and cephamycin biosynthesis. In: Fiechter A (ed) Adv in Biochem Eng Biotech. Springer, Berlin Heidelberg New York

3. Lancini G, Lorenzetti R (1993) Biotechnology of antibiotics and other bioactive microbial metabolites, Plenum Press, New York
4. Skatrud PL, Tietz AJ, Ingolia TD, Cantwell CA, Fisher DK, Chapman JL, Queener SW (1989) Bio/Technol 7: 477
5. Aharonowitz Y, Cohen G, Martin JF (1992) Penicillin and cephalosporin biosynthetic genes: structure, organisation, regulation, and evolution. In: Orston LN, Balows A, Greenberg EP (eds) Ann Rev of Microbiol. Annual Review, Washington DC
6. Vandamme EJ (1990) In: Klein kauf H, von Dohren H (eds) Biochemistry of Peptide antibiotics: Recent advances in the biotechnology of -lactams and microbial bioactive peptides, Walter De Gruyter, Berlin
7. Shewale JG, Deshpande BS, Sudhakaron VK, Ambedkar SS (1990) Proc Biochem. June: 97
8. Hyun CK, Kim JH, Ryu DDY (1993) Biotech Bioeng 42: 800
9. Leinn J (1988) Eur Pat 0293218
10. Aramori I, Fukugawa M, Tsumara M, Iwami M, Ono H, Ishitani Y, Kojo H, Kohsaka M, Ueda Y, Iomanak, H (1992) J Ferment Bioeng 73: 185
11. Vincenzi JT, Hansen GJ (1993) Enzyme Microb Technol 15: 281
12. Szwajcer E, Miller J, Kovacevic S, Mosbach K (1990) Biochem Int 20: 1169
13. Binder R, Brown J, Romanick G (1994) Appl Environ Microbiol 60: 1805
14. Ishii Y, Saito K, Fujimura T, Isogai T, Kojo H, Yamashita M, Niwa M, Koshsaka M (1994) J Ferment Bioeng 77: 591
15. Totsuka K, Shimizu K, Konishi M, Yamamoto S (1992) Antimicrob Agents Chemother 36: 757
16. Takimoto A, Mitsushima K, Yagi S, Sonoyama T (1994) J Ferment Bioeng 77: 17
17. Park YH, Kim EY, Seo WT, Junj KH, Yoo YJ (1989) J Ferment Bioeng 67: 409
18. Lee H-J, Sohn Y-S, Ahn D-H, Kim H-S, Hyun HH (1992) Kor J Appl Microbiol Biotechnol 20: 178
19. Yung W-Y, Lin C-D, Chu I-M, Lee C-J (1994) Biotech Bioeng 43: 439
20. Dosondil M, Kleigen HH, Kolheb R, Mueller H, Newman A, Schugeri K (1990) Processtechnologie 6: 35
21. Kalyanpur M, Skea S, Siwak M (1985) In: Developments in Industrial Microbiology. Soc Indust Microbiol. Washington DC
22. Grabosch M (1987) Bioengg 1: 62
23. Kula MR (1985) Recovery Operations. In: Rehm H-J, Reed G, Braner H (eds) Biotechnology Vol. 6, VCH, Weinheim
24. Harrison STL (1991) Biotechn Adv 9: 217
25. Flaschel E, Frichs K (1993) Biotech Adv 11: 31
26. Baldwin JE, Shian CY, Byford MF, Schofield CJ (1994) Biochem J 30: 367
27. Gutierrez S, Diez B, Alvarij E, Barredo JL, Martin JF (1991) J Bacteriol 173: 2354
28. Jensen SE, Wong A, Rollins MJ, Westlake DWS (1990) J Bacteriol 172: 7269
29. Zhang JY, Demain Al (1990) Biotech Lett 12: 649
30. McCabe AP, Van Liempt H, Palissa H, Unkles SE, Riach MBR, Pfeifer E, von Dohren H, Kinghorn JR (1991) J Biol Chem 266: 12646
31. Hollander IJ, Shen Y-Q, Heim J, Demain AL (1984) Science 224: 610
32. Perry D, Abraham EP, Baldwin JE (1988) Biochem J 255: 345
33. Baldwin JE, Coates JB, Moloney MG, Pratt AJ, Wills AC (1990) Biochem J 266: 561
34. Jensen SE, Leskiew BK, Vining LC, Aharonowitz A, Westlake DWS, Wolfe S (1986) Can J Microbiol 32: 953
35. Castro JM, Liras P, Laiz L, Cortes J, Martin JF (1988) J Gen Microbiol 134: 133
36. Weigel BJ, Burgett SG, Chen VJ, Skatrud PL, Frolik CA, Queener SW, Ingolia TD (1988) J Bacteriol 170: 3817
37. Lubbe C, Wolfe S, Demain AL (1986) Appl Microb Biotechnol 23: 367
38. Kovacevic S, Tobin MB, Miller JR (1990) J Bacteriol 172: 3952
39. Laiz L, Liras P, Castro JM, Martin JF (1990) J Gen Microbiol 136: 663
40. Baldwin AE, Adlington RM, Coates B, Crabbe MJC, Crouch NP (1987) Biochem J 245: 831
41. Dotzlaf JE, Yeh WK (1989) J Biol Chem 264: 10219
42. Kovacevic S, Miller JR (1991) J Bacteriol 173: 398
43. Rollins MJ, Westlake DWS, Wolfe S, Jensen SE (1988) Can J Microbiol 34: 1196
44. Baker BJ, Dotzlaf JE, Yeh WK (1991) J Biochem 266: 5087
45. Cortes J, Martin JF, Castro JM, Laiz L, Liras PJ (1987) J Gen Microbiol 133: 3165
46. Felix HR, Nuesch J, Wehrli W (1980) FEMS Microbiol Lett 8: 55

47. Zhang J, Yuan J, Chen J, Phen Y (1993) Zhangguo Kangshenngsu Zazhi 18: 439
48. Isaacson DM, Kirschbaum (1986) Assay of Antimicrobiol substances In: Demain Al, Solomon NA (eds) Mannuals of Industrial Microbiology and Biotechnology, Amer Soc of Microbiol, Washington DC
49. Belter PA (1985) In: Moo-Young M, Cooney CL, Humphrey AE (eds), Comprehensive Biotechnology, Vol. 2, Pergamon Press, Oxford. pp. 473–480
50. Voser W (1982) J Chem Tech Biotechnol 32: 109
51. Chaubel MV, Payne GF, Reynolds CH, Alberts RL (1995) Biotech Bioeng 47: 215
52. Wildfenur ME (1985) Approach to Cephalosporin C purification In: Le Roith D, Shiloach J, Leaby TJ (eds) Purification of fermentation products: Application to large scale processes, American Chemical Society, Washington
53. Aracil MJ, Casillas L, Martinez RM (1993) Span Pat ES 2042, 383
54. Itagaki K, Ito T, Tejima H, Wada S (1993) Japan Pat 05, 230067.
55. Asai K, Nakamura F, Miyawaki T, Itoh K (1992) Japan Pat Appl 92/323051
56. Andrisano R, Guerra G, Mascellani G (1976) J Appl Chem Biotechnol 26: 459
57. Firoujtale E, Maikner JJ, Diessler CK, Cartier GP (1994) J Chromat 658: 361
58. Tanaka N, Araki M (1989) Polymer based packing materials for reverse phase liquid chromatography. In: Giddings JC, Grushka E, Brown PR (eds) Adv Chromat, 31, Marcel Dekker, New York
59. Lim CK (1992) Porous graphite carbon in biomedical application. In: Giddings JC, Grushka E, Brown PR (eds) Adv Chromat 32, Marcel Dekker, New York
60. Joseph JM (1986) ACS Symp Ser 297: 83
61. Schugerl K (1994) Solvent extraction in biotechnology: Recovery of primary and secondary metabolites, Springer Verlag, Berlin
62. Elk J (1977) In: Elk J (eds) Recent advances in the chemistry of beta-lactam antibiotics, Spec publ No. 28, The Chemical Society, London
63. Bywood R, Robinson C, Stables SC, Walker D, Wilson EM (1977) In: Elk J (eds) Recent advances in the chemistry of beta-lactam antibiotics. Spl pub No. 28, The Chemical Society, London
64. Hano T, Matsumoto M, Ohtake T, Hori F (1992) J Chem Eng Japan 25: 293
65. Lee H, Lee WK (1994) Sep Sci Technol 29: 601
66. Likidis Z, Schugerl K (1987) J Biotechnol 5: 293
67. Likidis, Schlichting E, Bischoff L, Schugerl K (1989) Biotech Bioeng 33: 1385
68. Harris TAJ, Khan S, Reuban GB, Shokaya T (1990) Reactive solvent extraction of beta-lactam antibiotics, In: Pyle DL (eds) Separations in Biotechnology, Elsevier, London
69. Borah MM, Dutta NN, Ghosh AC (1996) Bioseparation (in press)
70. Ho WSW, Li NN (1992) Emulsion liquid Membrane, In: Ho MSW, Sirkar KK (eds) Membrane Handbook, van Nostrand Reinhold, New York
71. Tsikas D, Scheper T, Schugerl K (1989) Chem Ing Tech 61: 418
72. Lee HK, Lee SC, Lee KW (1994) J Chem Tech Biotech 59: 365
73. Scheper T, Likidis Z, Makryaleas NC, Schugerl K (1987) Enzyme Microb Technol 9: 625
74. Sahoo GC, Dutta NN, Ghosh AC (1996) J Memb Sci 112: 147
75. Lee JC, Yeh JH, Yang WJ, Kan RC (1993) Biotech Bioeng 42: 527
76. Lee JC, Yeh JH, Kan RC (1994) Biotech Bioeng 43: 309
77. Tsikas D, Kaltsidon SE, Brunner G (1992) Chem Ing Tech, 64: 545
78. Ghosh AC, Borthakur S, Dutta NN (1995) Sep Tech 5(2) : 121
79. Kiani A, Bhave RR, Sirkar KK (1984) J Memb Sci 20: 125
80. Orchard JCA, Balls RP (1993) 3rd Int Conf Effective Membrane Process: New Perspectives, Bath, UK (May 12–14)
81. Basu R, Sirkar KK (1992) J Memb Sci 75: 131
82. Yang FZ, Rindfleisch D, Scheper T, Schugerl K (1993) 3rd Int Conf Effective Membrane process: New perspectives, Bath, UK (May 12–14)
83. Raghuraman B, Wiencek J (1993) AIChEJ 39: 1885
84. Boyadzhiev, Attanassova L (1994) Process Biochem 29: 237
85. Heath CA, Belfort G (1992) Synthetic membranes in Biotechnology: Realities and Possibilities, In: Fiechter A (ed) Adv in Biochem Engg/Biotechnol 47: 45, Springer Verlag, Berlin
86. Ishimura F, Suga K (1992) Biotech Bioeng 34: 171
87. Greco G, Veronese F, Largajolli R, Giantredo L (1983) Eur J Appl Microbiol Biotechnol 18: 333

88. Gravett DP, Molnar TE (1986) Recovery of an extracellular antibiotics by ultrafiltration, In: McGregor WC (ed) Bioprocess technology, 1 Membrane Separation in Biotechnology, Marcel Dekker, New York
89. Hustedt H, Kroner KH, Kula MR (1985) Application of phase partitioning in biotechnology, In: Walter H, Brooks DE, Fisher D (eds) Partitioning in aqueous two phase systems, Academic Press, New York
90. Daimond AD, Hsu JT (1992) Aqueous two-phase system for biomolecule separation, In: Fiechter A (ed) Adv Biochem Eng/Biotechnol, 47, Springer, Berlin
91. Yang WY, Chu I-M (1990) Biotech Bioeng 41: 191
92. Paquet V, Myint M, Roque C, Soncaille P (1994) Biotech Bioeng 44: 445
93. Lee CK, Sandler SI (1990) J Ferment Bioeng 35: 408
94. Ohno M, Otsaka M, Yagisawa M, Kondo S, Oppinger H, Hoffman H, Sukatsch D, Hepner L, Male C (1985) Antibiotics, In: Gerhard W (ed) Ullmann's Encyclopaedia of Industrial Chemistry, VCH, Weinheimm
95. Cardoss JP (1993) Biotech Bioeng 42: 1068
96. Gupta MN, Mathiasson B (1994) Chem Ind, 5 Sept 673
97. Cantwell AM, Calderone R, Sienko M (1994) J Chromat, 316: 133
98. Adamovics AJ (1990) High performance liquid chromatography, In: Adamovics AJ (ed) Chromatographic analysis of Pharmaceuticals, 49, Marcel Dekker, New York
99. Quintens Z, Eykens J, Roets E, Hoogmartens J (1993) J Planar Chromat, 6: 181
100. Kirchner JG (1990) Thin layer chromatography: Techniques of Chemistry, Vol. 24, John Wiley and Sons, New York
101. Marelli LP (1972) Analytical procedures for cephalosporins, In: Flynn EH (ed) Cephalosporins and penicillins, Academic Press, New York
102. Gooden WE, Du Y, Ahluwalia R, Day RA (1994) Anal Lett 27: 2153
103. Sorel RHA, Hulshoff A (1983) Dynamic anion-exchange chromatography, In: Giddings JC, Grushka E, Cazes J, Brown PR (eds) Adv Chromat, Vol. 21, Marcel Dekker, New York
104. Sacco D, Dellacherie E (1983) J Liq Chromat 6: 2543
105. White ER, Corroll MA, Zaremobo JE (1977) J Antibiot 30: 811
106. White ER, Fox M (1982) J Antibiot 53: 1538
107. Ueno H, Nishikawa M, Suzuki S, Murakami M (1983) J Chromat 258: 262
108. Usher JJ, Lewis M, Hughes DW (1985) Anal Biochem, 149: 105
109. Brisson AM, Fourtillan JB (1981) J Chromat 223: 393
110. Barbato F, Grieco C, Silipo C, Vittoria A (1979) J Antiobiot, 34: 233
111. Kennedy JH (1978) J Chromat Sci 16: 492
112. Serenvirante AK, Jaywardane AL, Gamberttoglio JG (1984) J Liq Chromat 17: 4157
113. Ivory CF (1988) Sep Sci Technol 23: 875
114. Zheng SN, Yanemoto T, Tadaki T (1991) J Chem Eng Japan 24: 471
115. Betina V (1983), In: Deyl Z, Chrambach A, Everaet FM, Prusik Z (eds) Electrophoresis, Part B Application, Elsevier, Amsterdam
116. Baird MHI (1991) The Can J Chem Eng 69: 1287
117. Hairiri MH, Ely JF, Mansoori GA (1989) Biotechnol 7: 686
118. Papamichael N, Borner B, Hustedt H (1991) J Chem Tech Biotech 50: 457
119. Diamond AD, Hsu JT (1989) Biotech Bioeng 34: 1000

Dynamical Modelling, Analysis, Monitoring and Control Design for Nonlinear Bioprocesses

D. Dochain[1] and M. Perrier[2]
[1] Maître de Recherches FNRS, Cesame, Université Catholique de Louvain, Bâtiment Euler, 4-6 avenue G. Lemaître, B-1348 Louvain-La-Neuve, Belgium
[2] Départment de Génie Chimique, Ecole Polytechnique de Montréal, Succursale "Centre Ville", CP 6079, Montréal H3C 3A7, Canada

1 Introduction. 150
2 General Dynamical Model 152
  2.1 Example 1: PHB Producing Process 153
  2.2 General Dynamical Model. 155
  2.3 Example 2: Anaerobic Digestion 155
  2.4 Example 3. Yeast Growth 157
  2.5 Example 4. Activated Sludge Process. 158
  2.6 Fixed Bed Reactors. 159
3 Dynamical Analysis of Stirred Tank Bioreactor Models. 161
  3.1 A Key State Transformation. 161
  3.2 Model Order Reduction 162
    3.2.1 Singular Perturbation Technique for Low Solubility Products 162
    3.2.2 A General Rule for Order Reduction 163
    3.2.3 Example: the Anaerobic Digestion 164
4 Monitoring of Bioprocesses. Part I: Asymptotic Observers 165
  4.1 Asymptotic Observers for Single Tank Bioprocesses. 166
    4.1.1 Theoretical Convergence of the Asymptotic Observer 168
    4.1.2 Example: PHB Producing Process. 169
    4.1.3 Implementation Aspects: Choice of the Sampling Period. 170
  4.2 Application to a PHB Producing Process. 171
  4.3 Asymptotic Observers for Fixed Bed Bioreactors 174
    4.3.1 Practical Implementation of the Asymptotic Observer 176
    4.3.2 Stability Properties of the Asymptotic Observer. 178
5 Monitoring of Bioprocesses. Part II: On-line Estimation of Reaction Rates 178
  5.1 Statement of the Estimation Problem 179
  5.2 Observer-Based Estimator 179
  5.3 Application to the Bakers's Yeast Fed-Batch Process 180
    5.3.1 Experimental Verification 183
6 Adaptive Linearizing Control of Bioprocesses. 185
  6.1 Design of the Adaptive Linearizing Controller 185
  6.2 Example 1: Anaerobic Digestion 187
  6.3 Example 2: Activated Sludge Process. 191
7 Conclusions. 194
8 References 195

This paper is a survey on methods which have been developed and applied in the field of dynamical modelling, analysis, monitoring and control design of bioprocesses over the past fifteen years. A key feature of the paper is to show how to incorporate the well-known knowledge about the dynamics of biochemical processes (basically, the reaction network and the material balances) in monitoring and control algorithms. These are moreover capable of dealing with the process uncertainty (in particular on the reaction kinetics) by introducing, for instance in the control algorithms, an adaptation scheme.

Advances in Biochemical Engineering
Biotechnology, Vol. 56
Managing Editor: Th. Scheper
© Springer-Verlag Berlin Heidelberg 1997

List of Symbols

A	bioreactor cross-section (m^2)
A_1, A_2	transformation matrices
a_0	mass transfer coefficient
C_a, C_b	transformation matrices
C_1, C_2	estimation and/or controller gains
C_S	saturation constant of the oxygen ($g\,l^{-1}$)
D	dilution rate (h^{-1})
D_{am}	axial dispersion coefficient ($m^2\,s^{-1}$)
e	observation error
f, g	state and input functions of the output dynamical equation
F	feedrate vector ($g\,l^{-1}h^{-1}$)
F_{in}, F_R, F_W	influent, recycle and waste flowrate ($l\,h^{-1}$)
h	subvector of the known reaction rates
H	known matrix multiplying the unknown part of the reaction rates
k	yield coefficient
K	yield coefficient matrix
$k_L a$	oxygen mass transfer coefficient (h^{-1})
N	number of process components
M	number of reactions
P_{sat}	saturation concentration of the product P
q	number of discrete spatial points in a fixed bed reactor
q_m	specific maintenance rate (h^{-1})
Q	gaseous outflow rate ($g\,l^{-1}h^{-1}$)
Q_{in}	gaseous oxygen inflow rate ($g\,l^{-1}h^{-1}$)
Q_{out}	gaseous oxygen outflow rate ($g\,l^{-1}h^{-1}$)
$\Delta Q_{O_2}\,(= Q_{in} - Q_{out})$	gaseous oxygen feedrate ($g\,l^{-1}h^{-1}$)
r	reaction rate ($g\,l^{-1}h^{-1}$)
R	rank of the yield coefficient matrix K
RQ	respiration quotient
S, X, C, N, E, P	substrate, biomass, dissolved oxygen, nitrogen, ethanol, synthesis product concentrations ($g\,l^{-1}$)
S_{in}	influent substrate concentration ($g\,l^{-1}$)
u	system input
V, V_S	reactor and settler volumes (m^3)
W	air flow rate ($g\,h^{-1}$)
X_R	recycled biomass concentration ($g\,l^{-1}$)
y	system output
y^*	desired output value
z	space variable (m)
z_i	spatial position

Greek letters

α	estimation error dynamics pole
ε	singular perturbation parameter
ξ	component concentration vector $(g\,l^{-1})$
ζ	auxilary variable $(g\,l^{-1})$
ρ	vector of the unknown part of the reaction rates
μ	specific growth rate (h^{-1})
μ_o, μ_r, μ_e	specific growth rates of the respiratory growth on glucose, the oxido-reductive growth on glucose, and the respiratory growth on ethanol, respectively (h^{-1})
ν	specific production rate (h^{-1})
Π	auxiliary perturbation variable $(P = \Pi P_{sat})$

Indices

d	discretized
fi	fixed
fl	flowing-through
in	inflow
k	known
out	outflow
R	respirative
RF	oxido-reductive
t	time index
u	unknown
y	output

1 Introduction

Industrial-scale biotechnological processes have progressed vigorously over the last decades. Generally speaking, the problems arising from the implementation of these processes are similar to those of more classical industrial processes and the need for monitoring systems and automatic control in order to optimize production efficiency, improve product quality or detect disturbances in process operation is obvious. Nevertheless, automatic control of industrial biotechnological processes is clearly developing very slowly. There are two main reasons for this:

1. The internal working and dynamics of these processes are as yet badly grasped and many problems of methodology in modelling remain to be solved. It is difficult to develop models taking into account the numerous factors which can influence the specific bacterial growth rate. The modelling effort is often tedious and requires a great number of experiments before producing a reliable model. Reproducibility of experiments is often uncertain due to the difficulty in obtaining the same environmental conditions. Moreover, as these processes involve living organisms, their dynamic behaviour is strongly non-linear and non-stationary. Model parameters cannot remain constant over a long period: they will vary, e.g. due to metabolic variations of biomass or to random and unobservable physiological or genetic modifications. It should also be noted that the lack of accuracy of the measurements often leads to identifiability problems.
2. Another essential difficulty lies in the absence, in most cases, of cheap and reliable instrumentation suited to real-time monitoring. To date, the market offers very few sensors capable of providing reliable on-line measurements of the biological and biochemical parameters required to implement high performance automatic control strategies. The main variables (i.e. biomass, substrate and synthesis product concentrations) generally need to be determined by laboratory analyses. The cost and duration of the analyses obviously limit the frequency of the measurements.

The classical monitoring and control methods do not prove to be efficient enough for tiding over these basic difficulties. Therefore, to reconstruct the state of the system from the only on-line available measurements and to control the biological variables (biomass, substrates or synthesis products), appropriate algorithms have to be developed. The efficiency of any monitoring or control system highly depends on the design of the control and monitoring techniques and the care taken in their design. Indeed, monitoring or control algorithms will prove to be efficient if they are able to incorporate the important well-known information on the process while being able to deal with the missing information (lack of on-line measurements, uncertainty with regard to the dynamics, ...) in a "robust" way, i.e. such that these missing information will not significantly detract from the control performance of the process. The present generalized use of computers make it possible and fairly easy to design and implement controllers which are more sophisticated (and susceptible of better performance) than the classical PID. These controllers may refer to quite complex theory (nonlinear control, adaptive control) but, as it will be shown, their structure and implementation may remain rather simple while including the key features of simple PID's.

In this paper, we shall show how to incorporate the well-known knowledge about the dynamics of biochemical processes (basically, the reaction network and the material balances) in monitoring and control algorithms which are moreover capable of dealing with the process uncertainty (in particular with regard to the reaction kinetics) by introducing, for instance, in the control algorithms, an adaptation scheme). Figure 1 shows a schematic view of a computer-controlled bioreactor.

A key feature of this paper is to emphasize the central role played by linear algebra, which appears to be a very efficient tool in the design of monitoring and control algorithms for apparently complex nonlinear systems like bioprocesses, via the use of fairly simple and standard algebraic manipulations. It is also important to draw the attention of the reader to the fact that these control systems are not just the object of academic research but are already used in several applications (see e.g. [1], [2]). Adaptive as well as non-adaptive linearizing control of bioreactors has been quite an active research area over the last decade. In addition to the works of the present authors and their coworkers, let us also mention as examples [3–8]. A key reference to this paper is the book by Bastin and Dochain ([1]) in which a deeper theoretical analysis of the many monitoring and control methodologies introduced here can be found.

The paper is organized as follows. In Sect. 2, we shall introduce the general dynamical model for stirred tank bioreactors and illustrate it with several practical bioprocess examples [intracellular production of biodegradable polymers (poly-β-hydroxybutyric acid (PHB)), anaerobic digestion, yeast process, activated sludge process]. We shall also show in this section how to extend the general dynamical model to multi-tank processes or to non-perfectly mixed reactors like fixed bed reactors. Section 3 will concentrate on the dynamical analysis of the stirred tank reactor models by introducing a key state transformation and a general methodology for reducing the order of the model. In Sects. 4 and 5, we shall concentrate on the design of software sensors, i.e. algorithms based on the general dynamical model to estimate on-line the unknown parameters (like the reaction rates) and the unmeasured components from the few available on-line measurements. Sect. 4 will deal with the design of asymptotic observers for the process components, and Sect. 5 will concentrate on the on-line estimation of specific growth rates. In both cases, the proposed methodologies will be illustrated by real-life results on a PHB and a yeast process, respectively. Finally, Sect. 6 will present the design of adaptive linearizing controllers for bioprocesses based on a reduced order model of the process. The methodology will be illustrated with the anaerobic digestion and the activated sludge examples.

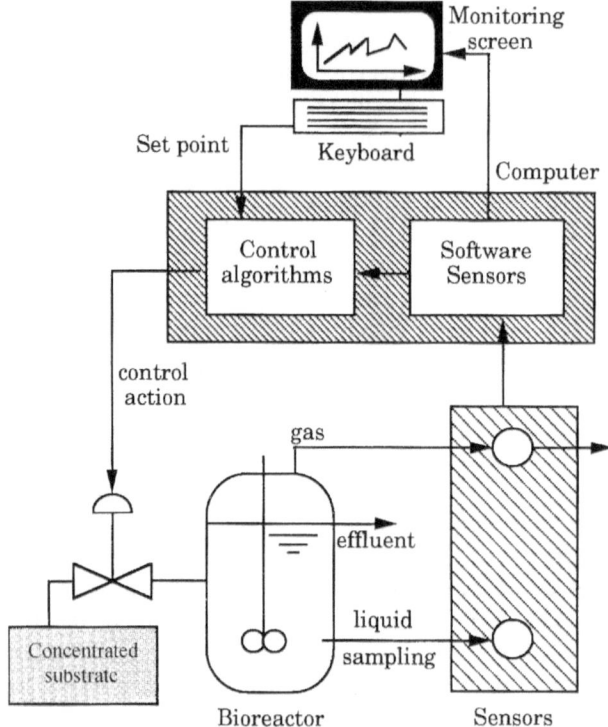

Fig. 1. Schematic view of a computer-controlled bioreactor

2 General Dynamical Model

A biotechnological process can be defined as a set of M biochemical reactions involving N components. The reactions most often encountered in bioprocesses are microbial growth (in which the biomass plays the role of an autocatalyst) and enzyme catalysed reactions (in which the biomass can be viewed as a simple catalyst); but many other reactions can also take place, like microorganism death, maintenance, etc. These reactions can be formalized into reaction schemes, as it is now illustrated. In the following we shall concentrate on four process examples which will be used in the rest of the text to illustrate the different themes of this paper:

1. a biodegrable polymer (PHB) production process;
2. yeast growth;
3. the two classical biological wastewater treatment processes: the anaerobic digestion and the activated sludge process.

The above choice is somewhat arbitrary (although the chosen processes correspond to typical processes which have been the object of research works over the past ten years and more, and for which estimation and/or control results are available). We could have considered indeed other process examples among which we would like to mention animal cell cultures and penicillin production. Let us mention to the interested reader some references concerning works related to the themes developed here: [9–11] on the animal cell cultures, and [12], [13] on penicillin G production.

2.1 Example 1: PHB Producing Process

Let us first consider the production of poly-β-hydroxybutyric acid (PHB), which is a biodegradable polymer. The PHB can be produced in an aerobic culture of *Alcaligenes eutrophus*, and the production may follow two metabolic pathways:

1. the first metabolic pathway for the production of PHB is growth associated and characterized by a very low yield. The growth takes place with three limiting substrates: oxygen, a source of carbon (e.g. fructose, glucose, acetic acid, methanol....) and nitrogen (usually under the form of ammonia)
2. the second metabolic pathway for the production of PHB is non-growth associated, where the biomass plays simply the role of a catalyst, and is completely inhibited by nitrogen.

In both cases, CO_2 is a by-product of the reaction. Therefore the production of PHB can be schematized by the following reaction network:

1) Growth associated production: $S + N + C \xrightarrow{\frown} X + P_1 + P_2$ \hfill (1)

2) Non-growth associated production: $S + C + X \longrightarrow X + P_1 + P_2$ \hfill (2)

where S, C, N, X, P_1 and P_2 represent the carbon source, the oxygen, the nitrogen, the biomass, the PHB and CO_2, respectively. In the first reaction scheme, the feedback arrow means that, in a growth reaction, the biomass is an autocatalyst, i.e. a product and a catalyst. The presence of X on both sides of the arrow in the second reaction scheme means that X simply plays the role of a catalyst.

Note that in the above reaction network the stoichiometric coefficients have been (deliberately) omitted, since in our view, each reaction scheme simply represents the qualitative mass exchange corresponding to the reaction. The objective of the reaction network here is to use it as a tool for deriving the dynamical model in a general manner. This will be explained in Sect. 2.2.

Note finally that in the following, the symbols used for the components (S, C, N, X, P_1, P_2 in the above example) will either designate the components in the reaction networks or their concentration ($g\,l^{-1}$) in the dynamical model.

Let us now concentrate on the derivation of the dynamical equations of the PHB process. By considering mass balances for each component in a stirred tank reactor (STR), we obtain the following equations:

$$\frac{dS}{dt} = -DS + DS_{in} - k_1 r_1 - k_2 r_2 \tag{3}$$

$$\frac{dC}{dt} = -DC + Q_{in} - Q_{out} - k_3 r_1 - k_4 r_2 \tag{4}$$

$$\frac{dN}{dt} = -DN + DN_{in} - k_5 r_1 \tag{5}$$

$$\frac{dX}{dt} = -DX + r_1 \tag{6}$$

$$\frac{dP_1}{dt} = -DP_1 + k_6 r_1 + r_2 \tag{7}$$

$$\frac{dP_2}{dt} = -DP_2 - Q_2 + k_7 r_1 + k_8 r_2 \tag{8}$$

D is the dilution rate (h^{-1}), S_{in} and N_{in} the influent carbon source and nitrogen concentrations (gl^{-1}). Q_{in} and Q_{out} are the inlet and outlet gaseous oxygen flowrates (gh^{-1}), and Q_2 is the CO_2 gaseous outflow rate (gl^{-1}). r_1 and r_2 the reaction rates (h^{-1}) of the reactions (1) and (2), respectively, and k_i ($i = 1$ to 8) the yield coefficients.

Note that each reaction rate (r_1 and r_2) has been normalized with respect to one component, the biomass concentration X and the PHB concentration P_1, respectively. Note also that the equations are valid whatever the operating conditions (continuous, fedbatch or batch). The above equations (3)–(8) can be rewritten in the following matrix form (in which we have set $\Delta Q_{0_2} = Q_{in} - Q_{out}$):

$$\frac{d}{dt}\begin{pmatrix} S \\ C \\ N \\ X \\ P_1 \\ P_2 \end{pmatrix} = -D\begin{pmatrix} S \\ C \\ N \\ X \\ P_1 \\ P_2 \end{pmatrix} + \begin{pmatrix} DS_{in} \\ \Delta Q_{02} \\ DN_{in} \\ 0 \\ 0 \\ 0 \end{pmatrix} - \begin{pmatrix} 0 \\ 0 \\ 0 \\ 0 \\ 0 \\ Q_2 \end{pmatrix}$$

$$+ \begin{pmatrix} -k_1 & -k_2 \\ -k_3 & -k_4 \\ -k_5 & 0 \\ 1 & 0 \\ k_6 & 1 \\ k_7 & k_8 \end{pmatrix} \begin{pmatrix} r_1 \\ r_2 \end{pmatrix} \tag{9}$$

2.2 General Dynamical Model

The dynamical model (9) can be rewritten in the following more compact form:

$$\frac{d\xi}{dt} = -D\xi + Kr + F - Q \tag{10}$$

where ξ is the vector of the bioprocess component ($\dim(\xi) = $ N), K is the yield coefficient matrix ($\dim(K) = $ N × M), r is the reaction rate vector ($\dim(r) = $ M), F is the feed rate vector and Q the gaseous outflow rate vector ($\dim(F) = \dim(Q) = $ N). The model (10) has been called the *General Dynamical Model* for stirred tank bioreactors (for further details on the notion scheme and the general dynamical model of bioreactors, see [1]). The derivation of the dynamical model from a reaction network is then straightforward by noting that each component k_{ij} of the yield coefficient matrix:

$$K = [k_{ij}] \quad i = 1 \text{ to } N, j = 1 \text{ to } M$$

is representative of the ith component: it is negative if the component is a reactant, it is positive if it is a product and it is equal to zero if the component does not intervene in the reaction.

2.3 Example 2: Anaerobic Digestion

Anaerobic digestion is a biological wastewater treatment process which produces methane. Four metabolic paths [14] can be identified in this process: two for acidogenesis and two for methanization (Fig. 2).

In the first acidogenic path (Path 1), glucose is decomposed into fatty volatile acids (acetate, propionate), hydrogen and inorganic carbon by acidogenic bacteria. In the second acidogenic path (Path 2), OHPA (Obligate Hydrogen Producing Acetogens) decompose propionate into acetate, hydrogen and inorganic carbon. In a first methanization path (Path 3), acetate is transformed into methane and inorganic carbon by acetoclastic methanogenic bacteria, while in the second methanization path (Path 4), hydrogen combines with inorganic carbon to produce methane under the action of hydrogenophilic methanogenic bacteria. The process can then be described by the following reaction network:

$$S_1 \xrightarrow{\quad} X_1 + S_2 + S_3 + S_4 + S_5 \tag{11}$$

$$S_2 \xrightarrow{\quad} X_2 + S_3 + S_4 + S_5 \tag{12}$$

$$S_3 \xrightarrow{\quad} X_3 + S_5 + P_1 \tag{13}$$

$$S_4 + S_5 \xrightarrow{\quad} X_4 + P_1 \tag{14}$$

where $S_1, S_2, S_3, S_4, S_5, X_1, X_2, X_3, X_4$, and P_1 are respectively glucose, propionate, acetate, hydrogen, inorganic carbon, acidogenic bacteria, OHPA (Obligate Hydrogen Producing Acetogens), acetoclastic methanogenic bacteria,

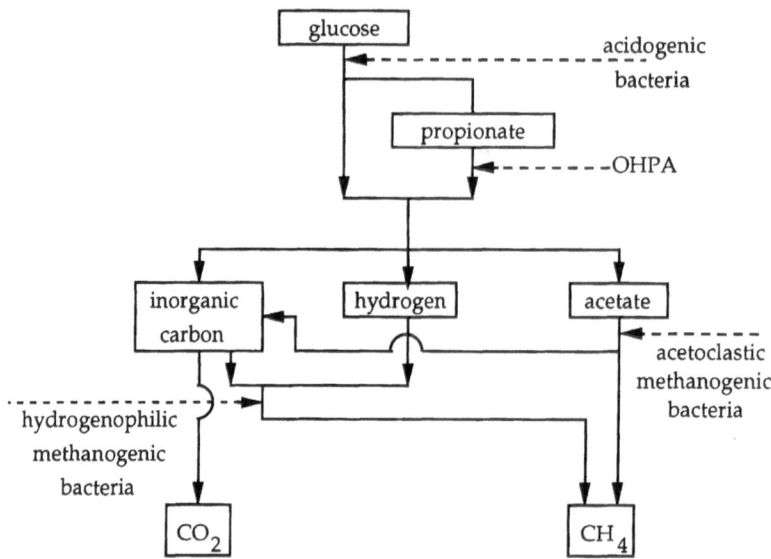

Fig. 2. Scheme of the anaerobic digestion

hydrogenophilic methanogenic bacteria and methane. The dynamical model of the anaerobic digestion process ($N = 10$, $M = 4$) in a stirred tank reactor can be described within the above formalism (10) by using the following definitions:

$$
\xi = \begin{pmatrix} X_1 \\ S_1 \\ X_2 \\ S_2 \\ X_3 \\ S_3 \\ X_4 \\ S_4 \\ S_5 \\ P_1 \end{pmatrix}, \quad
K = \begin{pmatrix}
1 & 0 & 0 & 0 \\
-k_{21} & 0 & 0 & 0 \\
0 & 1 & 0 & 0 \\
k_{41} & -k_{42} & 0 & 0 \\
0 & 0 & 1 & 0 \\
k_{61} & k_{62} & 0 & -k_{63} \\
0 & 0 & 0 & 1 \\
k_{81} & k_{82} & 0 & -k_{84} \\
k_{91} & k_{92} & k_{93} & -k_{94} \\
0 & 0 & k_{03} & k_{04}
\end{pmatrix} \tag{15}
$$

$$
F = \begin{pmatrix} 0 \\ D\,S_{in} \\ 0 \\ 0 \\ 0 \\ 0 \\ 0 \\ 0 \\ 0 \\ 0 \end{pmatrix}, \quad
Q = \begin{pmatrix} 0 \\ 0 \\ 0 \\ 0 \\ 0 \\ 0 \\ 0 \\ Q_1 \\ Q_2 \\ Q_3 \end{pmatrix}, \quad
r = \begin{pmatrix} r_1 \\ r_2 \\ r_3 \\ r_4 \end{pmatrix} = \begin{pmatrix} \mu_1 X_1 \\ \mu_2 X_2 \\ \mu_3 X_3 \\ \mu_4 X_4 \end{pmatrix} \tag{16}
$$

where μ_1, μ_2, μ_3, μ_4 are the specific growth rates (h^{-1}) of reactions (11), (12), (13), (14), respectively, and S_{in}, Q_1, Q_2 and Q_3 represent respectively the influent glucose concentration (gl^{-1}) and the gaseous outflow rates (gl^{-1}h^{-1}) of H_2, CO_2 and CH_4.

2.4 Example 3. Yeast Growth

Yeast (*Saccharomyces cerevisiae*) growth is usually characterized by the following three metabolic reactions (see e.g. [15]):

1. respiratory growth on glucose;
2. reductive growth on glucose;
3. respiratory growth on ethanol.

These can be formalized by the following reaction schemes:

$$S + C \xrightarrow{} X + P \tag{17}$$

$$S \xrightarrow{} X + E + P \tag{18}$$

$$E + C \xrightarrow{} X + P \tag{19}$$

where S, C, X, P and E are, respectively, glucose, oxygen, yeast, carbon dioxide and ethanol. In absence of growth, substrate and oxygen may be consumed for maintenance, which may be formalized by the following reaction scheme:

$$S + C \longrightarrow P \tag{20}$$

The dynamical model of the yeast growth (N = 5, M = 4) deduced from material balances can be formalized within the general dynamical model framework (10) by considering the following definitions:

$$\xi = \begin{pmatrix} S \\ C \\ X \\ P \\ E \end{pmatrix}, \quad K = \begin{pmatrix} -k_1 & -k_2 & 0 & -k_{10} \\ -k_3 & 0 & -k_4 & -1 \\ 1 & 1 & 1 & 0 \\ k_5 & k_6 & k_7 & k_{11} \\ 0 & k_8 & -k_9 & 0 \end{pmatrix} \tag{21}$$

$$r = \begin{pmatrix} r_1 \\ r_2 \\ r_3 \\ r_4 \end{pmatrix} = \begin{pmatrix} \mu_o X \\ \mu_r X \\ \mu_e X \\ q_m X \end{pmatrix}, \quad F = \begin{pmatrix} D S_{in} \\ \Delta Q_{o_2} \\ 0 \\ 0 \\ 0 \end{pmatrix}, \quad Q = \begin{pmatrix} 0 \\ 0 \\ 0 \\ Q_1 \\ 0 \end{pmatrix} \tag{22}$$

where μ_o, μ_r and μ_e are the specific growth rates of the respiratory growth on glucose (17), the reductive growth on glucose (18) and the respiratory growth on ethanol (19), respectively (h^{-1}), q_m is the specific maintenance rate (h^{-1}), k_i (i = 1 to 9) are yield coefficients, S_{in} is the influent glucose concentration (gl^{-1}), ΔQ_{o_2} the gaseous oxygen feed rate and Q_1 the gaseous outflow rate of carbon dioxide P (gl^{-1}h^{-1}).

2.5 Example 4. Activated Sludge Process

In the preceding examples, D is a scalar, but it may be a vector if there are more than one tank, as it is illustrated in the activated sludge process example.

The activated sludge process is one other classical biological (but aerobic) wastewater treatment process. It is usually operated in two sequential tanks (see Fig. 3): an aerator (in which the degradation of the pollutants S takes place) and a settler (which is used to recycle part of the biomass X to the aerator). The reaction in the aerator is usually described by a simple microbial growth (see e.g. [16], [17]).

$$S + C \xrightarrow{\quad} X \tag{23}$$

while the dynamics of the settler are described by the following mass balance equation:

$$\frac{dX_R}{dt} = \frac{F_{in} + F_R}{V_S} X - \frac{F_R + F_W}{V_S} X_R \tag{24}$$

where X_R is the concentration of the recycled biomass $(g\,l^{-1})$, F_{in}, F_R and F_W are the influent, recycle and waste flow rates $(g\,l^{-1}\,h^{-1})$, respectively, and V_S the settler volume (l). By considering the volume V (l) of the aerator and defining

$$D_{in} = \frac{F_{in}}{V}, \qquad D_2 = \frac{F_R}{V}, \qquad D_1 = D_{in} + D_2,$$

$$D_3 = \frac{F_{in} + F_R}{V_S}, \qquad D_4 = \frac{F_R + F_W}{V_S} \tag{25}$$

The dynamical equations of the process $(N = 4, M = 1)$ can be rewritten in the formalism of the general dynamical model (10) with the following definitions:

$$\xi = \begin{pmatrix} S \\ C \\ X \\ X_R \end{pmatrix}, \quad K = \begin{pmatrix} -k_1 \\ -kl_2 \\ 1 \\ 0 \end{pmatrix}, \quad F = \begin{pmatrix} D_{in} S_{in} \\ \Delta Q_{O_2} \\ 0 \\ 0 \end{pmatrix}, \quad r = \mu X, \quad Q = 0 \tag{26}$$

$$D = \begin{pmatrix} D_1 & 0 & 0 & 0 \\ 0 & D_2 & 0 & 0 \\ 0 & 0 & D_1 & -D_2 \\ 0 & 0 & -D_3 & D_4 \end{pmatrix} \tag{27}$$

Note that D is now a matrix. In activated sludge processes, the oxygen feed rate ΔQ_{O_2} term in the dynamical equation of the dissolved oxygen is usually set equal to the liquid-gas oxygen transfer rate:

$$\Delta Q_{O_2} = k_L a (C_S - C) \tag{28}$$

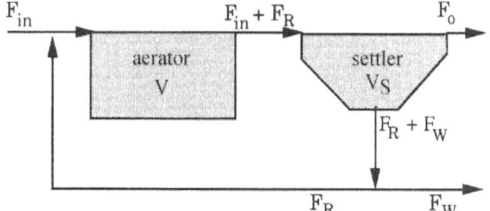

Fig. 3. Schematic view of an activated sludge process

where $k_L a$ is the oxygen mass transfer coefficient and C_S the saturation constant. In line with [18], we shall also consider in the following that $k_L a$ is a linear function of the air flow rate W:

$$k_L a = a_0 W, \ a_0 > 0 \tag{29}$$

2.6 Fixed Bed Reactors

Let us now see how to extend the General Dynamical Model (10) to non-completely mixed reactors. As an example, we shall concentrate on fixed bed reactors with axial diffusion.

Since the reactor is no longer in completely mixed conditions, the mass balance has to be computed on a thin section dz of the reactor (and not on the whole volume of the reactor anymore) (see Fig. 4).

Let us assume that among the N process components, N_{fi} are micro-organisms entrapped or fixed on some support, and N_{fl} other reactants (essentially substrates and products) flow through the reactor. For simplicity, we also consider the cross-section of the bioreactor to be constant and equal to A. From mass balance considerations on a section dz, we can deduce the following dynamical model:

$$\frac{\partial \xi_{fi}}{\partial t} = K_{fi} r(\xi_{fi}, \xi_{fl}) \tag{30}$$

$$\frac{\partial \xi_{fl}}{\partial t} = -\frac{F_{in}}{A} \frac{\partial \xi_{fl}}{\partial z} + D_{am} \frac{\partial^2 \xi_{fl}}{\partial z^2} + K_{fl} r(\xi_{fl}, \xi_{fl}) \tag{31}$$

with the following Danckwerts [19] boundary conditions:

$$D_{am} \frac{\partial \xi_{fl}}{\partial z} = -\frac{F_{in}}{A} (\xi_{fl, in} - \xi_{fl}) \quad \text{at } z = 0 \tag{32}$$

$$\frac{\partial \xi_{fl}}{\partial z} = 0 \quad \text{at } z = H \tag{33}$$

In the above equations, ξ_{fi} is the biomass concentration vector (dim $\xi_{fi} = N_{fi}$), ξ_{fl} is the other reactant concentration vector (dim $\xi_{fl} = N_{fl}$), $\xi_{fl, in}(t)$ is the influent concentration of ξ_{fl} (which is different from zero only for external substrates), $r(\xi_{fi}, \xi_{fl})$ is the reaction rate vector

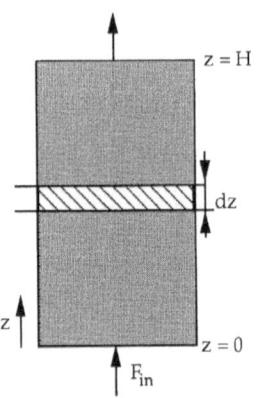

Fig. 4. Schematic view of a fixed bed reactor

(dim $r = M$), K_{fi} and K_{fl} are the yield coefficient matrices (dim $K_{fi} = N_{fi} \times M$; dim $K_{fl} = N_{fl} \times M$), F_{in} the hydraulic flow rate (m^3 h^{-1}), D_{am} is the axial dispersion coefficient (m^2 s^{-1}) and z is the space variable (m) ($z \in]0, H]$).

The dynamics of the bioprocess are described by partial differential equations. Yet note the similarity between this model (30), (31) and the General Dynamical Model (10) of stirred tank reactors. An important difference lies in the hydrodynamics term (the partial derivative term

$$-\frac{F_{in}}{A}\frac{\partial \xi_{fl}}{\partial z} + D_{am}\frac{\partial^2 \xi_{fl}}{\partial z^2}$$

instead of $- D\xi + F - Q$) and in the presence of boundary conditions (32), (33).

Remark 1: Note that the dynamical model of the fixed bed reactor in absence of diffusion, i.e. of the plug flow reactor, is readily obtained from the (30), (31), (32) by simply setting the dispersion coefficient D_{am} to zero ($D_{am} = 0$). Since there is then only a first order derivative of the state variable ξ_{fl} with respect to the space variable z, only the boundary condition at the reactor input ($z = 0$) is kept (32) and the boundary condition at the reactor output (33) is dropped.

Remark 2: The Danckwerts boundary conditions (32), (33) are largely accepted in chemical engineering, and found to be rather accurate in many applications. However, these conditions, and particularly the output reactor one (33), are sometimes the object of criticism (e.g. [20]). In particular, they are based on steady-state arguments, and (33) is strictly valid only for a reactor of infinite length.

Comment: The rest of the paper will mainly concentrate on stirred tank reactors. In order to keep the paper within a reasonable length, only Sect. 4 will deal with non-completely mixed reactors: it will be shown how to extend the asymptotic observer design to fixed bed reactors. Let us suggest that the interested reader look at the following references which deal with the dynamical modelling, analysis and control design for distributed parameter reactors [21], the adaptive linearizing control for fixed bed reactors [22], [23] and fluidized bed reactors [24], and with the approximation of fixed bed reactor models [25] and the control analysis [26] via singular perturbation techniques.

3 Dynamical Analysis of Stirred Tank Bioreactor Models

3.1 A Key State Transformation

The key result of this section is the use of a state transformation by which part of the dynamical model (10) becomes independent of the reaction kinetics r (see [1], [27]). This transformation will play a very important role in the design of asymptotic observers (Sect. 4) and adaptive linearizing controllers (Sect. 6). The proposed transformation indeed readily derives from the notion of invariants in reaction systems (see e.g. [28], [29]).

The transformation is defined as follows. Let us denote the rank of the matrix K by R (rank(K) = R), and consider the following partition of the process component vector:

$$\xi = \begin{bmatrix} \xi_a \\ \xi_b \end{bmatrix} \tag{34}$$

where ξ_a contains R (arbitrarily chosen) process variables and ξ_b the others, but such that the corresponding submatrix K_a is full rank (rank(K_a) = R). Let us define the state transformation ζ (dim(ζ) = N − R):

$$\zeta = C_a \xi_a + C_b \xi_b \tag{35}$$

where C_a and C_b are solutions of the matrix equation:

$$C_a K_a + C_b K_b = 0 \tag{36}$$

In the particular (but quite general) situation of independent irreversible reactions, then R = M and C_b may be chosen to be a full rank square matrix. Let us derive the dynamics of ζ from (10) and the definition (35).

1) *Single reactor: D = scalar*

$$\frac{d\zeta}{dt} = -D\zeta + C_a(F_a - Q_a) + C_b(F_b - Q_b) \tag{37}$$

2) *Multi-reactor: D = matrix*

$$\frac{d\zeta}{dt} = -(C_b D_{bb} + C_a D_{ab})C_b^{-1}\zeta + C_a(F_a - Q_a) + C_b(F_b - Q_b)$$

$$+ [(C_b D_{bb} + C_a D_{ab})C_b^{-1} - C_b D_{ba} - C_a D_{aa}]\xi_a \tag{38}$$

with:

$$D = \begin{pmatrix} D_{aa} & D_{ab} \\ D_{ba} & D_{bb} \end{pmatrix} \tag{39}$$

3) *Fixed bed reactor*

For simplicity reasons, let us consider here $C_b = I$ and let us put the vector ξ_{fi} of the fixed components in ξ_b:

$$\xi_b = \begin{pmatrix} \xi_{bf} \\ \xi_{fi} \end{pmatrix} \tag{40}$$

Then we can rewrite the auxiliary variable ζ as follows:

$$\zeta = \begin{pmatrix} \zeta_{fl} \\ \zeta_{fi} \end{pmatrix} = \begin{pmatrix} \xi_{bf} \\ \xi_{fi} \end{pmatrix} + \begin{pmatrix} C_{af} \\ C_{ae} \end{pmatrix} \xi_a \tag{41}$$

The dynamics of ζ can then be written as follows:

$$\frac{\partial \zeta_{fl}}{\partial t} = -\frac{F_{in}}{A} \frac{\partial \zeta_{fl}}{\partial z} + D_{am} \frac{\partial^2 \zeta_{fl}}{\partial z^2} \tag{42}$$

$$\frac{\partial \zeta_{fi}}{\partial t} = -\frac{F_{in}}{A} C_{ae} \frac{\partial \xi_a}{\partial z} + D_{am} C_{ae} \frac{\partial^2 \xi_a}{\partial z^2} \tag{43}$$

Note that the dynamical equations of ζ (37), (38) and (42), (43) are independent of the reaction kinetics $r(\xi)$.

3.2 Model Order Reduction

The examples of bioprocesses presented in the preceding sections have shown that a bioreactor dynamical model may be fairly complex in some instances and involve a large number of differential equations. But there are many practical applications where a simplified reduced order model is sufficient from an engineering viewpoint. Model simplification can be achieved by using the singular perturbation technique, which is a technique for transforming a set of $n + m$ differential equations into a set of n differential equations and a set of m algebraic equations. This technique is suitable when neglecting the dynamics of substrates and of products with low solubility in the liquid phase. The method will be illustrated with one specific example (low solubility product) before stating the general rule for order reduction.

3.2.1 Singular Perturbation Technique for Low Solubility Products

Let us consider a biochemical reaction described by the following reaction scheme

$$S \rightarrow P \tag{44}$$

where P is a volatile product which can be given off in gaseous form and has low solubility in the liquid phase. The dynamical model is as follows:

$$\frac{dS}{dt} = -r - DS + D S_{in} \tag{45}$$

$$\frac{dP}{dt} = kr - DP - Q \tag{46}$$

The consistency of this model requires that the product concentration P be related to a saturation concentration representative of the product solubility, which is expressed as:

$$P = \Pi P_{sat} \tag{47}$$

where P_{sat} is the saturation concentration which is constant in a stable physico-chemical environment. The model (45), (46) is rewritten in the standard singular perturbation form, with $\varepsilon = P_{sat}$:

$$\frac{dS}{dt} = -r - DS + DS_{in} \tag{48}$$

$$\varepsilon \frac{d\Pi}{dt} = kr - \varepsilon D \Pi - Q \tag{49}$$

If the solubility is very low, we obtain a reduced order model by setting $\varepsilon = 0$ and replacing the differential equation (49) by the algebraic one:

$$Q = kr \tag{50}$$

3.2.2 A General Rule for Order Reduction

The above example shows that the rule for model simplification is actually very simple and that an explicit singular perturbation analysis is not really needed. Consider that, for some i, the dynamics of the component ξ_i are to be neglected. The dynamics of ξ_i are described by Eq. (10):

$$\frac{d\xi_i}{dt} = -D\xi_i + K_i r + F_i - Q_i \tag{51}$$

where K_i is the line of K corresponding to the component ξ_i. The simplification is then achieved by setting ξ_i and $d\xi_i/dt$ to zero, i.e. by replacing the differential equation (51) by the following algebraic equation:

$$K_i r = -F_i + Q_i \tag{52}$$

It has been shown that the above model order reduction rule is not only valid for low solubility products but also for bioprocesses with fast and slow reactions. Then the above order reduction rule (52) applies to substrates of fast reactions (as long as they intervene only in fast reactions) (see [30] for further details).

Note the close connection between the singular perturbation reduction and the quasi-steady-state (QSS) approximation, which is largely used in (bio) chemical engineering. This suggests the following comment: singular perturbation can be viewed as an efficient mathematical tool to rigourously justify QSS approximations on a systematic basis via an appropriate analysis (including the choice of an appropriate *small* perturbation parameter).

3.2.3 Example: the Anaerobic Digestion

Let us see how to apply the above model order reduction rule (52) to a specific example, the anaerobic digestion. First of all, it is well-known that methane is a low solubility product. Therefore the above procedure applies. Furthermore, assume that the second methanization path (hydrogen consumption) is limiting, i.e. that the first three reactions (11), (12), (13) are fast and the fourth one (14) is slow. We can then apply the model order reduction rule (52) to the glucose concentration S_1, the propionate concentration S_2, the acetate concentration S_3 and the dissolved methane concentration P_1. By setting their values and their time derivatives to zero:

$$S_1 = S_2 = S_3 = P_1, \qquad \frac{dS_1}{dt} = \frac{dS_2}{dt} = \frac{dS_3}{dt} = \frac{dP_1}{dt} = 0 \qquad (53)$$

we reduce their differential equations to the following set of algebraic equations:

$$\begin{pmatrix} -k_{21} & 0 & 0 & 0 \\ k_{41} & -k_{42} & 0 & 0 \\ k_{61} & k_{62} & -k_{63} & 0 \\ 0 & 0 & k_{03} & k_{04} \end{pmatrix} \begin{pmatrix} r_1 \\ r_2 \\ r_3 \\ r_4 \end{pmatrix} = \begin{pmatrix} -D S_{in} \\ 0 \\ 0 \\ Q_3 \end{pmatrix} \qquad (54)$$

By inverting the submatrix of the yield coefficients of the left-hand side of (54), we can express the reaction rates r_1, r_2, r_3 and r_4 as functions of the feedrate DS_{in} and of the gaseous methane outflow rate Q_3:

$$r_1 = \frac{1}{k_{21}} D S_{in} \qquad (55)$$

$$r_2 = \frac{k_{41}}{k_{21} k_{42}} D S_{in} \qquad (56)$$

$$r_3 = \frac{k_{41} k_{62} + k_{42} k_{61}}{k_{21} k_{42} k_{63}} D S_{in} \qquad (57)$$

$$r_4 = \frac{1}{k_{04}} Q_3 - \frac{k_{03}}{k_{04}} \frac{k_{41} k_{62} + k_{42} k_{61}}{k_{21} k_{42} k_{63}} D S_{in} \qquad (58)$$

Let us replace the reaction rates r_1, r_2 and r_4 by their above expressions (55), (56), (58) in the dynamical equation of the hydrogen concentration S_4, which is then rewritten as follows:

$$\frac{dS_4}{dt} = -DS_4 - Q_1 - k_1 Q_3 + k_2 D S_{in} \tag{59}$$

where k_1 and k_2 are defined as follows:

$$k_1 = \frac{k_{84}}{k_{04}}, \qquad k_2 = \frac{k_{81}}{k_{21}} + \frac{k_{41} k_{82}}{k_{21} k_{42}} + \frac{k_{03} k_{84}}{k_{04}} \frac{k_{41} k_{62} + k_{42} k_{61}}{k_{21} k_{42} k_{63}} \tag{60}$$

Equation (59) will be used for the design of an adaptive linearizing controller of the hydrogen concentration in Sect. 5. Note that the coefficients k_1 and k_2 are nonlinear combinations of the yield coefficients k_{ij}.

4 Monitoring of Bioprocesses. Part I: Asymptotic Observers

A key question in bioprocess control is how to monitor reactant and product concentrations in a reliable and cost effective manner. However, it appears that, in many practical applications, only some of the concentrations of the components involved and critical for quality control are available for on-line measurement. For instance, dissolved oxygen concentration and gaseous flow rates are available for on-line measurement while the values of the concentrations of biomass, substrates and/or synthesis products are often available via off-line analysis. An interesting alternative which circumvents and exploits the use of a model in conjunction with a limited set of measurements are the use of Luenberger or Kalman observers. In these techniques, a model, which includes states that are measured as well as states that are not measured, is used in parallel with the process and the model states may then be used for feedback. This configuration may be used to reduce the effect of noise on measurements as well as to reconstruct the states that are not measured. An introduction to these ideas can be found in e.g. [31]. These concepts were originally developed for linear problems. Because of the nonlinear characteristics of the bioprocess dynamics, it is of interest to extend these concepts and exploit particular structures for biochemical engineering application problems. Linearized versions (the linearized tangent model) of the process dynamics are computed from a Taylor's series expansion of a state space model around some equilibrium point and the observer theory referred to above can be applied. These modified observers, particularly the extended Kalman filter (EKF), has found applications in some biochemical processes (e.g. [32], [33], [34], [35]).

One of the reasons for the popularity of the EKF is that it is easy to implement since the algorithm can be derived directly from the state space model. However, since (as the extended Luenberger observer) it is based on a linearized model of the process, the stability and convergence properties are essentially local and valid around some equilibrium point, and it is rather difficult to guarantee its stability over wide ranges of operation. [36] shows that the EKF for state and parameter estimation of linear systems may give biased estimates or even diverge if it is not carefully initialized. Let us also point out that the derivation of the EKF is based on some stochastic assumptions on the measurement and process noises, which might be questionable in practice.

One reason for the problem of convergence of EKF is that, in order to guarantee the (arbitrarily chosen) exponential convergence of the observer, the process must be locally observable, i.e. the linearized tangent model must be observable and fulfill the classical observability rank condition. This condition, as it turns out, is restrictive in many practical situations and may account for the failure of EKF to find widespread application (e.g. [1], [37], [38]).

Another problem is that the theory for the extended Luenberger and Kalman observers is developed using a perfect knowledge of the system parameters, in particular of the process kinetics: it is difficult to develop error bounds and there is often a large uncertainty on these parameters.

It appears from the above remarks that there is a clear incentive to develop new methodologies for the on-line estimation of the unmeasured concentration variables in biochemical reaction systems that do not rely on the explicit use of kinetic models. Indeed, the objective of this section is to propose an alternative to EKF and use process physics in a more direct manner to develop a nonlinear observer applicable to the estimation problem of stirred tank reactors (STR). The proposed observer is based on the well-known nonlinear model of the process without the knowledge of the process kinetics being necessary. In order to advance the application of this method, we discuss its stability and convergence properties. We would like to emphasize that the presented results are global (i.e. independent of the initial conditions) as opposed to the local properties for EKF. The approach (called *asymptotic observers*) proved to be very successful when applied to bioreactors (see e.g. [1]). The proposed asymptotic observer for STR's is an intermediate method between the "classical" observers (EKF or extended Luenberger observer), which require a full process model knowledge, and the adaptive observers ([39], [40]), which include state and parameter estimation within the same estimation scheme. A review of the application of adaptive observers to biochemical processes is given by Dochain [41].

This section is organized as follows. We shall first present the general methodology for single tank bioprocesses and discuss its theoretical convergence properties and the practical implementation aspects. Then we shall present a real-life application on a PHB producing process. Finally we shall introduce the extension to fixed bed reactors.

4.1 Asymptotic Observers for Single Tank Bioprocesses

The derivation of the asymptotic observer equations are based on the key state transformation (35), (36) introduced in Sect. 3.1 and on the following assumptions:

1. M components are measured on-line[1].
2. The feedrates F, the gaseous outflow rates Q and the dilution rate D are known either by measurement or by choice of the user.
3. The yield coefficient matrix K is known.
4. The reaction rate vector r is unknown.
5. The M reactions are irreversible and independent, i.e. $\text{rank}(K) = R = M$

From Assumption 1, we can define the following state partition:

$$\xi = \begin{bmatrix} \xi_1 \\ \xi_2 \end{bmatrix} \tag{61}$$

where ξ_1 and ξ_2 hold for the measured component concentrations and the unmeasured ones, respectively.

Let us consider one (arbitrarily chosen) transformation ζ, defined by (35) and (36). The variable ζ can be rewritten as a linear combination of the measured and unmeasured states ξ_1 and ξ_2, i.e.:

$$\zeta = C_a \xi_a + C_b \xi_b = A_1 \xi_1 + A_2 \xi_2 \tag{62}$$

Recall that the dynamics of ζ are independent of the reaction rate $r(\xi)$:

$$\frac{d\zeta}{dt} = -D\zeta + C_a(F_a - Q_a) + C_b(F_b - Q_b) \tag{63}$$

Equations (62) and (63) are the basis for the derivation of the asymptotic observer. The dynamical equations of ζ are used to calculate an estimate of ζ on-line, which is used, via Eq. (62) and the on-line data of ξ_1, to derive an estimate the unmeasured component ξ_2. Let us further assume that the matrix A_2 is invertible. Then the asymptotic observer is written as follows:

$$\frac{d\hat{\zeta}}{dt} = -D\hat{\zeta} + C_a(F_a - Q_a) + C_b(F_b - Q_b) \tag{64}$$

$$\hat{\xi}_2 = A_2^{-1}[\hat{\zeta} - A_1 \xi_1] \tag{65}$$

Comment: If we consider the most simple and straightforward choice for the state transformation ζ, i.e. with $\xi_1 = \xi_a$ and $\xi_2 = \xi_b$, and with $C_b = I_{N-p}$ (i.e. the identity matrix of order $N - p$), then we have:

$$A_2 = I_{N-p}, \qquad A_1 = -K_2 K_1^{-1} \tag{66}$$

And therefore the condition on the invertibility of A_2 is in fact a condition on the invertibility of the submatrix K_1 (i.e. K_1 is full rank or rank $(K_1) = M$).

The observer (64), (65) is completely independent of the process kinetics and can be implemented without the knowledge of the reaction rates $r(\xi)$ being required.

---

[1] The asymptotic observer, when the number of measured components is larger than M, is developed and discussed in [1].

4.1.1 Theoretical Convergence of the Asymptotic Observer

The convergence properties can summarized in the following theorem.

Theorem 4. *If the dilution rate D is a persistently exciting signal, i.e. if there exist positive constants δ and β such that:*

$$\delta = \int_{t}^{t+\beta} D(\tau)d\tau \tag{67}$$

then:

$$\lim_{t \to \infty} (\xi_2 - \hat{\xi}_2) = 0 \tag{68}$$

Proof. The proof of the theorem is immediate if one observes that, from (37), (64), (65), the dynamics of the estimation error is equal to:

$$\frac{d(\xi_2 - \hat{\xi}_2)}{dt} = -D(\xi_2 - \hat{\xi}_2) \tag{69}$$

QED

Remark 1: The persistence-of-excitation condition on D simply requires that D is not equal to zero for too long. This condition is clearly easily fulfilled in fedbatch and continuous reactors.

Remark 2: Note that the stability of the asymptotic observer only depends on the dilution rate and not on the kinetics. In other words, the reactor may be unstable (due to the kinetics in the Haldane inhibition model) while the asymptotic observer is asymptotically stable (because of stable hydrodynamics).

Remark 3: The asymptotic observer is derived under the assumption that the yield coefficients are known and constant. This assumption is indeed very important, since an important preliminary step in the design of the observer is to consider a reliable reaction network of the process. This means that the chosen reaction network should be at the same time representative of the process and simple enough (in order to limit the number of required on-line measurements). "Representative" means that it reflects the major reaction mechanisms of the process (e.g. at the exclusion of side reactions that have a minor influence). In such a case, the yield coefficients should remain rather constant (ideally they would be constant by definition of the representativity of the reaction network). The yield coefficients considered for the asymptotic observer are then determined via preliminary (e.g. batch) experiments and/or stoichiometric arguments. If the yield coefficients are expected to change during the course of the process operation, our suggestion is to periodically re-evaluate them via the use of off-analyses combined to the on-line data.

Remark 4 (reversible reactions): Note that, in presence of reversible reactions, the matrix K will not be full column rank because it will contain two identical columns. However a simple way to treat the asymptotic observation problem of reversible reactions is to consider each reversible reaction as *one global reaction* (whose rate may then be positive or negative) and therefore characterized by *only one* column in the matrix. This means that if the "forward" and "backward" reactions are characterized by a reaction rate r_1 and r_2 respectively, we consider, for the observation, one global reaction characterized by the same stoichiometric coefficients but one global reaction $r_3 = r_1 - r_2$.

4.1.2 Example: PHB Producing Process

Let us first define one state transformation ζ, e.g.:

$$\xi_a = \begin{bmatrix} X \\ P_1 \end{bmatrix}, \qquad \xi_b = \begin{bmatrix} S \\ C \\ N \\ P_2 \end{bmatrix} \tag{70}$$

with:

$$K_a = \begin{bmatrix} 1 & 0 \\ k_6 & 1 \end{bmatrix}, \qquad K_b = \begin{bmatrix} -k_1 & -k_2 \\ -k_3 & -k_4 \\ -k_5 & 0 \\ k_7 & k_8 \end{bmatrix} \tag{71}$$

Therefore if C_b is chosen as an identity matrix $(C_b = I_4)$, then C_a is equal to:

$$C_a = -K_b K_a^{-1} = \begin{bmatrix} k_1 - k_2 k_6 & k_2 \\ k_3 - k_4 k_6 & k_4 \\ k_5 & 0 \\ -k_7 + k_6 k_8 & -k_8 \end{bmatrix} \tag{72}$$

The dynamics of ζ are here equal to:

$$\frac{d}{dt} \begin{pmatrix} \zeta_1 \\ \zeta_2 \\ \zeta_3 \\ \zeta_4 \end{pmatrix} = -D \begin{pmatrix} \zeta_1 \\ \zeta_2 \\ \zeta_3 \\ \zeta_4 \end{pmatrix} + \begin{pmatrix} DS_{in} \\ \Delta Q_{02} \\ DN_{in} \\ -Q_2 \end{pmatrix} \tag{73}$$

Note that Since $F_a = Q_a = 0$, the dynamics of ζ are also independent of the yield coefficients.

Case 1: X and P_1 are measured on-line
Then $A_1 = C_a$ and $A_2 = C_b = I_4$. Therefore, S, C, N and P_2 can be estimated by using the asymptotic observer via the dynamical equation of ζ (73) as follows:

$$\hat{S} = \hat{\zeta}_1 + (k_2 k_6 - k_1)X - k_2 P_1 \tag{74}$$

$$\hat{C} = \hat{\zeta}_2 + (k_4 k_6 - k_3)X - k_4 P_1 \tag{75}$$

$$\hat{N} = \hat{\zeta}_3 - k_5 X \tag{76}$$

$$P_2 = \hat{\zeta}_4 + (k_7 - k_6 k_8)X + k_8 P_1 \tag{77}$$

Case 2: C and P_2 are measured on-line
This choice corresponds to the components which are probably a priori the easiest ones to be measured on-line. Then we have a different state partition for ξ_1 and ξ_2 than the one used for ζ:

$$\xi_1 = \begin{bmatrix} C \\ P_2 \end{bmatrix}, \qquad \xi_2 = \begin{bmatrix} S \\ N \\ X \\ P_1 \end{bmatrix} \tag{78}$$

and A_1 and A_2 are then equal to:

$$A_1 = \begin{pmatrix} 0 & 0 \\ 1 & 0 \\ 0 & 0 \\ 0 & 1 \end{pmatrix}, \qquad A_2 = \begin{pmatrix} 1 & 0 & k_1 - k_2 k_6 & k_2 \\ 0 & 0 & k_3 - k_4 k_6 & k_4 \\ 0 & 1 & k_5 & 0 \\ 0 & 0 & -k_7 + k_6 k_8 & -k_8 \end{pmatrix} \tag{79}$$

It is straightforward to check that the matrix A_2 is invertible if $k_3 k_8 \neq k_4 k_7$. In practice, this means that the respiratory quotient must be different from 1 (RQ $\neq 1$). But the situation RQ $= 1$ is the one usually encountered in PHB producing processes (for further details, see [1]). This example shows that not any choice of M measured components is valid for the implementation of the asymptotic observer: the submatrix K_1 must be full rank, this means that the measured components must independent (here, C and P_2 are not independent if RQ $= 1$) or the measured components have to take part in all the reactions (at least one in each reaction) in order to avoid a submatrix K_1 with (a) column(s) filled with zeros (A_2 will not be full rank if $\xi_1 = [NX]^T$, for instance).

4.1.3 Implementation Aspects: Choice of the Sampling Period

Stability of the discrete-time asymptotic observer

Practical computer implementation of the asymptotic observer (64), (65) requires that it be rewritten in a discrete-time form. This can be done simply by replacing the time derivative of ζ by a finite difference (using a first order Euler approximation):

$$\frac{d\hat{\zeta}}{dt} \rightarrow \frac{\hat{\zeta}_{t+1} - \hat{\zeta}_t}{T} \tag{80}$$

where T is the sampling period and t and $t + 1$ are time indices. The asymptotic observer is then written as follows for the general case $p \geq M$:

$$\hat{\zeta}_{t+1} = \hat{\zeta}_t - TD_t \hat{\zeta}_t + TC_a(F_{at} - Q_{at}) + C_b(F_{bt} - Q_{bt}) \tag{81}$$

$$\hat{\xi}_{2,t+1} = A_2^{-1} [\hat{\zeta}_{t+1} - A_1 \xi_{1,t+1}] \tag{82}$$

For the discrete-time equation, the value of the sampling period plays a role in the stability. In fact, if the dilution rate D is bounded as follows:

$$0 \leq D(t) \leq D_{max} \tag{83}$$

then Eq. (81) will remain stable as long as T is smaller than $2/D_{max}$:

$$T \leq \frac{2}{D_{max}} \tag{84}$$

This can be shown by considering the following positive definite decrescent Lyapunov function:

$$W_t = (\zeta_t - \hat{\zeta}_t)^T (\zeta_t - \hat{\zeta}_t) \tag{85}$$

the time difference of which is equal to:

$$W_{t+1} - W_t = [TD_t(TD_t - 2)](\zeta_t - \hat{\zeta}_t)^T (\zeta_t - \hat{\zeta}_t) \tag{86}$$

which is nonpositive definite, decrescent as long as inequality (84) holds.

Dynamics of the discrete-time asymptotic observer

Equation (84) gives a condition for the stability of the discrete-time version of the asymptotic observer. But even if the sampling period T is chosen so as to fulfill condition (84), large values of T may introduce oscillations in the estimation. As a matter of fact, if we assume that D is constant, the dynamics of ζ is characterized by a (discrete-time) pole equal to $(1 - TD)$. It will be negative if T is larger than $1/D(T > 1/D)$ and then correspond to an oscillating (underdamped) dynamics for ζ. Therefore, in order to avoid (undesirable) oscillations, condition (84) can be replaced by:

$$T \leq \frac{1}{D_{max}} \tag{87}$$

Remark: Note that different sampling periods may be used for the computation of the variables ζ and the calculation of the observation of ξ_2, e.g. a "fast" computation of the variables ζ and a slower computation of the estimated values of ξ_2. This choice may depend on the measurement sampling interval which may be different from one variable to another.

4.2 Application to a PHB-Producing Process

One interesting aspect of this example is that it shows that by using extra process information, the monitoring algorithm can be simplified. Indeed we have shown here above that because the respiratory quotient is equal to one ($RQ = 1$), the measurements of oxygen and carbon dioxide are not independent and cannot be used to implement the asymptotic observer for the two-reaction PHB process.

However we have not yet used the fact that the non-growth associated production of PHB is completely inhibited by nitrogen, which on the other hand is a limiting substrate for growth. And in practice the process is operated in two sequential steps: a first step for growth *in presence of nitrogen*, and a second step for non-growth associated production *in the absence of nitrogen*. Therefore the process can be viewed as a sequence of two reactions and not simply a two-reaction process. This means that just one measured component is necessary to reconstruct the state of the bioreactor. Here we had chosen oxygen whose data appeared to be more reliable. Therefore we designed an asymptotic observer for each step. The first step in the design is the rewriting of the process dynamics, which are indeed the dynamical model (9), but once with $r_2 = 0$, and the other time with $r_1 = 0$ and $N = 0$.

1) *Growth without production*

$$\frac{d}{dt}\begin{vmatrix} S \\ C \\ N \\ X \\ P_1 \\ P_2 \end{vmatrix} = -D\begin{vmatrix} S \\ C \\ N \\ X \\ P_1 \\ P_2 \end{vmatrix} + \begin{vmatrix} D\,S_{in} \\ \Delta Q_{o2} \\ D\,N_{in} \\ 0 \\ 0 \\ 0 \end{vmatrix} - \begin{vmatrix} 0 \\ 0 \\ 0 \\ 0 \\ 0 \\ Q_2 \end{vmatrix} + \begin{vmatrix} -k_1 \\ -k_3 \\ -k_5 \\ 1 \\ k_6 \\ k_7 \end{vmatrix} r_1 \qquad (88)$$

2) *Production without growth*

$$\frac{d}{dt}\begin{vmatrix} S \\ C \\ X \\ P_1 \\ P_2 \end{vmatrix} = -D\begin{vmatrix} S \\ C \\ X \\ P_1 \\ P_2 \end{vmatrix} + \begin{vmatrix} D\,S_{in} \\ \Delta Q_{o2} \\ 0 \\ 0 \\ 0 \end{vmatrix} - \begin{vmatrix} 0 \\ 0 \\ 0 \\ 0 \\ Q_2 \end{vmatrix} + \begin{vmatrix} -k_2 \\ -k_4 \\ 0 \\ 1 \\ k_8 \end{vmatrix} r_2 \qquad (89)$$

Let us concentrate on the observation of the fructose concentrations S, the nitrogen concentration N, the biomass concentration X and the PHB concentration P_1 from the measurement of the dissolved oxygen C.

Let us consider the following variable ζ for both steps:

$$\zeta = \xi_2 - K_2 K_1^{-1} \xi_1 \qquad (90)$$

which corresponds to the following choices of vectors and matrices:

$$\xi_a = \xi_1, \qquad \xi_b = \xi_2, \qquad C_b = 1, \qquad C_a = -K_2 K_1^{-1} \qquad (91)$$

This means that for each step, we have:

1) *Growth without production*

$$\xi_1 = C, \quad \xi_2 = \begin{pmatrix} S \\ N \\ X \\ P_1 \end{pmatrix} \tag{92}$$

$$\zeta_1 = S - \frac{k_1}{k_3} C, \quad \zeta_2 = N - \frac{k_5}{k_3} C, \quad \zeta_3 = X + \frac{1}{k_3} C, \quad \zeta_4 = P_1 + \frac{k_6}{k_3} C \tag{93}$$

2) *Production without growth*

$$\xi_1 = C, \quad \xi_2 = \begin{pmatrix} S \\ X \\ P_1 \end{pmatrix} \tag{94}$$

$$\zeta_5 = S - \frac{k_2}{k_4} C, \quad \zeta_6 = X, \quad \zeta_7 = P_1 + \frac{1}{k_4} C \tag{95}$$

Since N is not measured on-line, the passage from step 1 to step 2 is based on the value of the estimate of the nitrogen concentration, $\hat{N}$, given by the asymptotic observer in step 1. The asymptotic observer in its discrete-time has then the following form:

1) *Growth without production: whenever* $\hat{N}_t > 0$

$$\hat{\zeta}_{1,t+1} = \hat{\zeta}_{1,t} - TD_t \hat{\zeta}_{1,t} + TD_t S_{in,t} - \frac{k_1}{k_3} T\Delta Q_{O2,t} \tag{96}$$

$$\hat{\zeta}_{2,t+1} = \hat{\zeta}_{2,t} - TD_t \hat{\zeta}_{2,t} + TD_t N_{in,t} - \frac{k_5}{k_3} T\Delta Q_{O2,t} \tag{97}$$

$$\hat{\zeta}_{3,t+1} = \hat{\zeta}_{3,t} - TD_t \hat{\zeta}_{3,t} + \frac{k_1}{k_3} T\Delta Q_{O2,t} \tag{98}$$

$$\hat{\zeta}_{4,t+1} = \hat{\zeta}_{4,t} - TD_t \hat{\zeta}_{4,t} + \frac{k_6}{k_3} T\Delta Q_{O2,t} \tag{99}$$

$$\hat{S}_{t+1} = \hat{\zeta}_{1,t+1} + \frac{k_1}{k_3} C_{t+1} \tag{100}$$

$$\hat{N}_{t+1} = \hat{\zeta}_{2,t+1} + \frac{k_5}{k_3} C_{t+1} \tag{101}$$

$$\hat{X}_{t+1} = \hat{\zeta}_{3,t+1} - \frac{1}{k_3} C_{t+1} \tag{102}$$

$$\hat{P}_{1,t+1} = \hat{\zeta}_{4,t+1} - \frac{k_6}{k_3} C_{t+1} \tag{103}$$

2) *Production without growth: when* $\hat{N}_t = 0$

$$\hat{\zeta}_{5,t+1} = \hat{\zeta}_{5,t} - TD_t\hat{\zeta}_{5,t} + TD_t\,S_{in,t} - \frac{k_2}{k_4}\,T\Delta Q_{O2,t} \tag{104}$$

$$\hat{\zeta}_{6,t+1} = \hat{\zeta}_{6,t} - TD_t\hat{\zeta}_{6,t} \tag{105}$$

$$\hat{\zeta}_{7,t+1} = \hat{\zeta}_{7,t} - TD_t\hat{\zeta}_{7,t} + \frac{1}{k_4}\,T\Delta Q_{O2,t} \tag{106}$$

$$\hat{S}_{t+1} = \hat{\zeta}_{5,t+1} + \frac{k_2}{k_4}\,C_{t+1} \tag{107}$$

$$\hat{X}_{t+1} = \hat{\zeta}_{6,t+1} \tag{108}$$

$$\hat{P}_{1,t+1} = \hat{\zeta}_{7,t+1} - \frac{1}{k_4}\,C_{t+1} \tag{109}$$

The above asymptotic observer has been implemented on a 20-liter pilot fedbatch reactor of the Solvay company, Belgium (see [42]). The values of the yield coefficients are equal to:

$$k_1 = 2.5,\ k_2 = 2.13,\ k_3 = 1.23,\ k_4 = 1,\ k_5 = 0.11,\ k_6 = 10 \tag{110}$$

Figure 5(a–d) shows the data of dissolved oxygen C, inlet and outlet oxygen gaseous flowrates $Q_{O2,in}$ and $Q_{O2,out}$, volume V, and inlet fructose and nitrogen concentration S_{in} and N_{in}. Figure 5(e–h) shows the estimation results for S, N, X and P: note the good correspondence of the software observation with the off-line analyses. Note that the switching from step 1 to step 2 is efficiently driven by the estimate of nitrogen $\hat{N}$.

Remark: The design of the above asymptotic observer can be easily extended to multi-reactor processes, since the definition of the transformation ζ is the same. The main question is the stability of the dynamics of the auxiliary variables ζ (38). We shall not develop this point here, which is largely discussed in [27]. We shall simply concentrate on one example (activated sludge process) in Sect. 6, and discuss its stability.

4.3 Asymptotic Observers for Fixed Bed Bioreactors

Let us now discuss the extension and application of the asymptotic observer design to non-completely mixed reactors. Let us consider here only the components which flow through the reactor, ξ_{fl}, to the exclusion of the fixed components, ξ_{fi}. In consequence, we shall only consider the subvector ζ_{fl} in the auxiliary variable ζ, defined in Eq. (41). Let us modify the first assumption introduced in Sect. 4.1:

1b. p (= M) components are measured on-line *along the reactor*.
and introduce a sixth one:

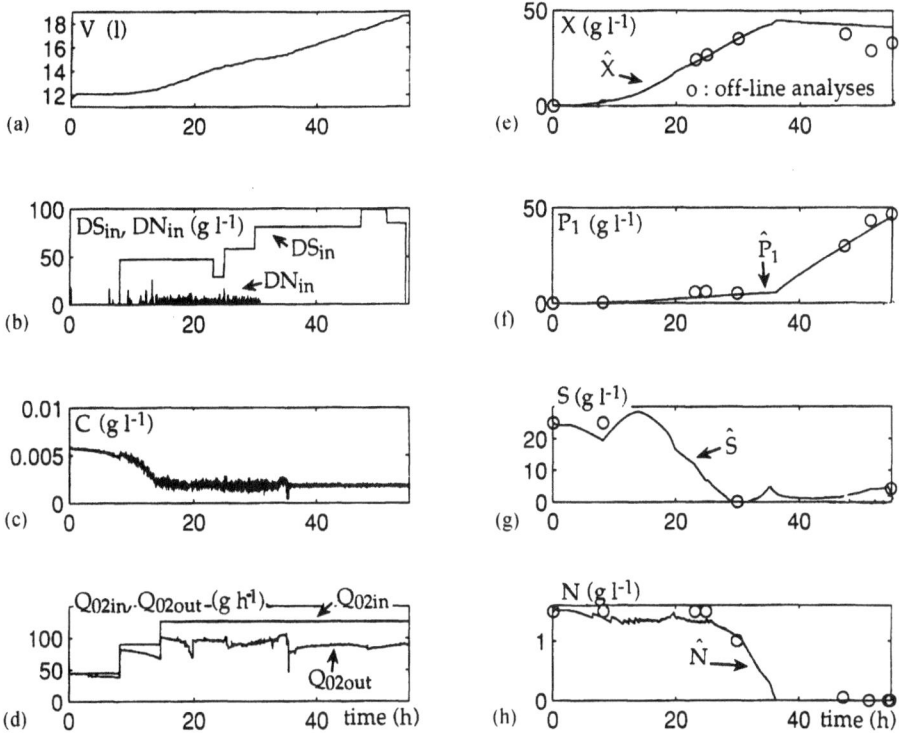

Fig. 5a–h. On-line data and estimation results of the asymptotic observer of the PHB process

6. The axial mass dispersion coefficient D_{am} and the reactor section A are known.

Assume that the measured components are put in the vector ξ_a (i.e. we make a particular choice for ξ_1 and ξ_2 ($\xi_1 = \xi_a$, $\xi_2 = \xi_{bf}$; this is done only to simplify the approach but of course other choices are still possible). Then by using the same arguments as in Sect. 4.1 and Eqs. (31), (41) and (42), we obtain the following asymptotic observer for fixed bed bioreactors:

$$\frac{\partial \hat{\zeta}_{fl}}{\partial t} = -\frac{F_{in}}{A}\frac{\partial \hat{\zeta}_{fl}}{\partial z} + D_{am}\frac{\partial^2 \hat{\zeta}_{fl}}{\partial z^2} \tag{111}$$

$$\hat{\xi}_2 = \hat{\zeta}_{fl} - C_{af}\xi_1 \tag{112}$$

There remain two key questions with the above asymptotic observer:

1) The above observer is written in the form of ("infinite dimensional") partial derivative equations (PDE's). These are not very easy nor convenient to handle

in practical control and monitoring applications. Moreover, it is assumed that M components are available for on-line measurement *along the reactor*. In line with a number of works on the subject (e.g. [43], [44]), we propose to "reduce" the above PDE equations to a finite number of ordinary differential equations (ODE's) at a finite number of positions along the reactor and to consider the reduced equations for the practical application of the observer. This will the subject of Sect. 4.3.1.

2) The second key question is to know whether the proposed observer is reliable, i.e. under which conditions will it give estimates that converge to their true values: this will be the subject of Sect. 4.3.2.

4.3.1 Practical Implementation of the Asymptotic Observer

Let us reduce the PDE's of the asymptotic observer (111) to a finite number of ODE's. We shall not discuss here the choice of one reduction method (see e.g. [45], [46] for this topic); we shall only assume that the user has chosen one method for approximating the PDE's (e.g. finite differences, orthogonal collocation,...) and that the reduced model is a fairly good representation of the PDE asymptotic observer (111). Whatever the reduction method, the partial derivatives of the variables ζ_i with respect to the space variable z are approximated by a weighted sum of z_i at a finite number of positions z_j ($j = 0$ to q, where 0 and q hold for the input and output of the reactor, respectively) along the reactor:

$$\frac{\partial^k}{\partial z^k} \begin{pmatrix} \zeta_i(z = z_1) \\ \zeta_i(z = z_2) \\ \vdots \\ \zeta_i(z = z_q) \end{pmatrix} \cong [\tilde{c}_k | \tilde{C}_k] \begin{pmatrix} \zeta_i(z = z_0) \\ \zeta_i(z = z_1) \\ \vdots \\ \zeta_i(z = z_q) \end{pmatrix}, \quad k = 1, 2 \tag{113}$$

$$dim(\tilde{c}_k) = q \times 1, \quad dim(\tilde{C}_k) = q \times q \tag{114}$$

This procedure results in a rewriting of the auxiliary variables ζ_{fl} and of the asymptotic observer equations under ODE's at each position z_j ($j = 1$ to q). Then the dynamics of the auxiliary variables ζ_{fl} and of the asymptotic observer become:

Dynamics of ζ_{fl}:

$$\frac{d\zeta_d}{dt} = \left[D_{am} C_2 - \frac{F_{in}}{A} C_1 \right] \zeta_d + \left[D_{am} c_2 - \frac{F_{in}}{A} c_1 \right] (\xi_{2,in} + C_{af} \zeta_{1,in}) \tag{115}$$

Asymptotic observer:

$$\frac{d\hat{\zeta}_d}{dt} = \left[D_{am} C_2 - \frac{F_{in}}{A} C_1 \right] \hat{\zeta}_d + \left[D_{am} c_2 - \frac{F_{in}}{A} c_1 \right] (\xi_{2,in} + C_{af} \zeta_{1,in}) \tag{116}$$

with:

$$\zeta_d = \begin{vmatrix} \zeta_i(z = z_1) \\ \vdots \\ \zeta_i\,(z = z_q) \\ \vdots \\ \zeta_{Nf-q}\,(z = z_1) \\ \vdots \\ \zeta_{Nf-q}(z = z_q) \end{vmatrix}, \qquad \xi_{f,in} = \begin{pmatrix} \xi_{1,in} \\ \xi_{2,in} \end{pmatrix} \tag{117}$$

$$\tilde{C}_k = \begin{pmatrix} \tilde{C}_k & 0 & \cdots & 0 \\ 0 & \tilde{C}_k & \cdots & \vdots \\ \vdots & \vdots & \cdots & 0 \\ 0 & 0 & \cdots & \tilde{C}_k \end{pmatrix}, \qquad \tilde{c}_k = \begin{pmatrix} \tilde{c}_k \\ \tilde{c}_k \\ \vdots \\ \tilde{c}_k \end{pmatrix}, \quad k = 1, 2 \tag{118}$$

One important feature of (115) and (116) is that the fixed bed reactor is indeed approximated by a stirred multi-tank reactor. In case of a finite difference approximation, the equations represent a cascade of stirred tank reactors; but with other approximation methods (e.g. orthogonal collocation), the model exhibits interconnections between each of the stirred tank reactors of the ODE model since the entries of the matrices $\tilde{C}_k$ ($k = 1, 2$) are generally different from zero.

A simple example

If we consider a finite difference approximation, the matrices $\tilde{C}_k$ ($k = 1, 2$) are equal to:

$$\tilde{C}_1 = \frac{1}{\Delta z} \begin{pmatrix} 1 & 0 & 0 & \cdots & 0 & 0 \\ -1 & 1 & 0 & \cdots & \vdots & \vdots \\ 0 & -1 & 1 & \cdots & \vdots & \vdots \\ 0 & 0 & -1 & \cdots & \vdots & \vdots \\ \vdots & \vdots & \vdots & \cdots & 1 & 0 \\ 0 & 0 & 0 & \cdots & -1 & 1 \end{pmatrix}, \qquad \tilde{c}_1 = \frac{1}{\Delta z} \begin{pmatrix} 1 \\ 0 \\ \vdots \\ 0 \end{pmatrix} \tag{119}$$

$$\tilde{C}_2 = \frac{1}{(\Delta z)^2} \begin{pmatrix} -2 & 1 & 0 & \cdots & 0 & 0 \\ 1 & -2 & 1 & \cdots & \vdots & \vdots \\ 0 & 1 & -2 & \cdots & \vdots & \vdots \\ 0 & 0 & 1 & \cdots & \vdots & \vdots \\ \vdots & \vdots & \vdots & \cdots & -2 & 1 \\ 0 & 0 & 0 & \cdots & 1 & -2 \end{pmatrix}, \qquad \tilde{c}_2 = \frac{1}{(\Delta z)^2} \begin{pmatrix} 1 \\ 0 \\ \vdots \\ 0 \end{pmatrix} \tag{120}$$

where Δz is the spatial discretization step.

4.3.2 Stability Properties of the Asymptotic Observer

Let us concentrate on the stability properties of the reduced form (116) of the asymptotic observer. If we define the observation error:

$$e = \zeta_d - \hat{\zeta}_d \tag{121}$$

then the dynamics of the observation error e is readily obtained from (115) and (116):

$$\frac{de}{dt} = \left[D_{am} \, C_2 - \frac{F_{in}}{A} \right] e \tag{122}$$

Therefore the stability depends on the state matrix:

$$D_{ma} \, C_2 - \frac{F_{in}}{A} \, C_1 \tag{123}$$

Because of the diagonal structure of the matrices C_1 and C_2, the stability of the above state matrix (123) depends on the stability of each submatrix:

$$D_{ma} \, \tilde{C}_2 - \frac{F_{in}}{A} \, \tilde{C}_1 \tag{124}$$

Therefore it follows that the asymptotic observer (116) will be asymptotically stable, i.e.:

$$\lim_{t \to \infty} \hat{\xi}_d = \hat{\xi}_d \tag{125}$$

if the eigenvalues of the matrix $D_{ma}\tilde{C}_2 - \frac{F_{in}}{A} \tilde{C}_1$ are stable. Note that here again the stability of the asymptotic observer only depends on the hydrodynamics and not on the kinetics.

Note also that it is routine to check the stability of the matrix by using any matrix computation program. An interesting particular case is the finite difference approximation of a fixed bed reactor without dispersion ($D_{am} = 0$). Indeed, the stability then simply depends on the matrix $\frac{F_{in}}{A} C_1$, and it is straightforward to check that the matrix $\frac{F_{in}}{A} C_1$ [see Eq. (119)] is stable as long as F_{in} is positive.

Remark: Note that it is possible to extend the asymptotic observer to the *fixed* components ξ_{fi}. An example is given in [22].

5 Monitoring of Bioprocess.
Part II: On-line Estimation of Reaction Rates

In this section, we address the problem of estimating the reaction rates from on-line knowledge of the process components (knowledge available either from measurements or from state estimation). The statement of the estimation problem is presented first followed by the development of an observer-based estimator. Finally, an example on the estimation of specific growth rates for bakers' yeast is described in detail.

5.1 Statement of the Estimation Problem

We consider a biotechnological process described by the General Dynamical Model (10). In line with Sect. 4, we assume that:

1. The feedrates F, the gaseous outflow rates Q and the dilution rate D are known either by measurement or by choice of the user.
2. The yield coefficient matrix K is known.
3. The process components ξ are known either by measurement or by estimation using an asymptotic observer (as described in Sect. 4.1).
4. The reaction rate vector $r(\xi)$ is partially unknown and written as follows:

$$r(\xi) = \begin{pmatrix} H(\xi)\rho(\xi) \\ h(\xi) \end{pmatrix} \tag{126}$$

where $H(\xi)$ is a diagonal matrix of known functions of the state and $\rho(\xi)$ a vector of unknown functions of ξ with dim $\rho(\xi) = n_u$. The known reaction rates are given by vector $h(\xi)$ with dim $h(\xi) = M - n_u$.

Using Eq. (126), the General Dynamical Model is rewritten as:

$$\frac{d\xi}{dt} = K_u H(\xi)\rho(\xi) + K_k h(\xi) - D\xi - Q + F \tag{127}$$

where K_u and K_k are matrices of yield coefficients associated with the unknown and known reaction rates respectively.

5.2 Observer-Based Estimator

A state observer form is used to provide on-line information for updating the estimate of $\rho(\xi)$. The estimation algorithm is written as follows.

$$\frac{d\hat{\xi}}{dt} = K_u H(\xi)\hat{\rho} + K_k h(\xi) - D\xi - Q + F - \Omega(\xi - \hat{\xi}) \tag{128}$$

$$\frac{d\hat{\rho}}{dt} = [K_u H(\xi)]^T \Gamma(\xi - \hat{\xi}) \tag{129}$$

The update of the parameter vector $\hat{\rho}$ is driven by the deviation $(\xi - \hat{\xi})$ which reflects the mismatch between $\hat{\rho}$ and ρ. The matrices Ω and Γ are tuning parameters for adjusting the rate of convergence of the algorithm. A common choice is:

$$\Omega = diag(-\omega_i), \qquad \Gamma = diag(-\gamma_j), \quad \omega_i, \gamma_j > 0 \tag{130}$$

With this choice, the stability of (128) and (129) is satisfied (see [1] for further details). The tuning procedure may be simplified if the state equations are first decoupled using the following transformation:

$$\Psi = K_u^{-1}\xi \tag{131}$$

Applying the estimation algorithm to the transformed state equations yields:

$$\frac{d\hat{\Psi}}{dt} = H\hat{\rho} + K_u^{-1}K_k h - D\Psi + K_u^{-1}(F - Q) - \Omega(\Psi - \hat{\Psi}) \tag{132}$$

$$\frac{d\hat{\rho}}{dt} = H\Gamma(\Psi - \hat{\Psi}) \tag{133}$$

5.3 Application to the Baker's Yeast Fed-Batch Process

A modified version of the process model given in Sect. 2.4 is considered. From a global point of view, the bakers' yeast fed-batch process can only be in an ethanol production regime or in an ethanol consumption regime. The process model is divided into two partial models to represent the two regimes. The first partial model corresponds to the ethanol production regime and is denoted as the oxido-reductive (RF) partial model:

$$\frac{d\xi}{dt} = K_{RF}\cdot r_{RF} - D\xi - Q + F \tag{134}$$

where

$$K_{RF} = \begin{pmatrix} -k_1 & -k_2 & -k_{10} \\ -k_3 & 0 & -1 \\ 1 & 1 & 0 \\ k_5 & k_6 & k_{11} \\ 0 & k_8 & 0 \end{pmatrix}, \quad r_{RF} = \begin{pmatrix} \mu_o X \\ \mu_r X \\ q_m X \end{pmatrix}$$

The second partial model, the respirative (R) partial model, corresponds to the ethanol consumption regime where oxidation of both glucose and ethanol may occur. The mass balance for this regime is written as:

$$\frac{d\xi}{dt} = K_R r_R - D\xi - Q + F \tag{135}$$

where

$$K_R = \begin{pmatrix} -k_1 & 0 & -k_{10} \\ -k_3 & -k_4 & -1 \\ 1 & 1 & 0 \\ k_5 & k_7 & k_{11} \\ 0 & -k_9 & 0 \end{pmatrix}, \quad r_R = \begin{pmatrix} \mu_o X \\ \mu_e X \\ q_m X \end{pmatrix}$$

The first step in the estimation procedure is to identify the measured variables. The concentration of ethanol, dissolved oxygen, and carbon dioxide are available for measurement. However, it was shown in [47] that those three measurements are not linearly independent. The two independent measured state variables selected are dissolved oxygen C and carbon dioxide P concentrations. Two specific growth rate estimation algorithms are needed, one for the estimation of $[\mu_o \ \mu_r]$ in the ethanol production regime and one for the estimation of $[\mu_o \ \mu_e]$ in the ethanol consumption regime. The maintenance coefficient q_m is assumed to be known [47]. Only the derivation for the ethanol production partial model is given below, the procedure being identical for the other partial model. The mass balance equations for the measured state variables are:

$$\frac{dC}{dt} = -DC - k_3 \mu_o X + q_m X + \Delta Q_{O2} \tag{136}$$

$$\frac{dP}{dt} = -DP + k_5 \mu_o X + k_6 \mu_r X + k_{11} X - Q_1 \tag{137}$$

The biomass concentration is unknown and will be estimated by an asymptotic observer as described later. With reference to Eq. (127), we can define the following vector and matrices:

$$K_u = \begin{pmatrix} -k_3 & 0 \\ k_5 & k_6 \end{pmatrix}, \qquad H = \begin{pmatrix} \hat{X} & 0 \\ 0 & \hat{X} \end{pmatrix}, \qquad K_k = \begin{pmatrix} -1 \\ k_{11} \end{pmatrix}$$

$$\xi_1 = \begin{pmatrix} C \\ P \end{pmatrix}, \qquad \rho = \begin{pmatrix} \mu_o \\ \mu_r \end{pmatrix}, \qquad h = [q_m \ \hat{X}]$$

A linear transformation is applied to the measured state variables (ξ_1) set to decouple the equations in term of the specific growth rates:

$$\Psi = K_u^{-1} \xi_1 \tag{138}$$

The matrices of tuning parameters Ω and Γ in Eqs. (132) and (133) are chosen as follows:

$$\Omega = -HC_1, \qquad \Gamma = C_2 \tag{139}$$

where C_1 and C_2 are diagonal matrices. The elements of these two matrices are chosen to ensure constant dynamics of the estimation error throughout the experiment. With C_1 and C_2 chosen as:

$$C_1 = \begin{pmatrix} C_{11} & 0 \\ 0 & C_{11} \end{pmatrix} = \begin{pmatrix} \frac{2\alpha}{\hat{X}} & 0 \\ 0 & \frac{2\alpha}{\hat{X}} \end{pmatrix}, \qquad C_2 = \begin{pmatrix} \frac{C_{11}^2}{4} & 0 \\ 0 & \frac{C_{11}^2}{4} \end{pmatrix} \tag{140}$$

the poles of the error dynamics are all located at $-\alpha$. A single tuning parameter (α) is thus needed.

The non-measured biomass concentration appears in the equations of the estimation algorithm. A biomass concentration observer is thus needed for the application of the estimation algorithm. First, the process model is partitioned into two subsets. The first subset includes the equations associated with the measured state variables ($\xi_1 = [C\,P]^T$), while the second subset is associated with the non-measured state variable ($\xi_2 = X$):

$$\frac{d\xi_1}{dt} = -D\xi_1 + K_u H\rho + K_k h + F_1 - Q_1 \tag{141}$$

$$\frac{d\xi_2}{dt} = -D\xi_2 + K_2 H\rho \tag{142}$$

Applying the following linear transformation between the unmeasured and measured state variables:

$$\zeta = \xi_2 - K_2 K_u^{-1} \xi_1 \tag{143}$$

we obtain:

$$\frac{d\zeta}{dt} = -(D + K_2 K_u^{-1} K_k)\zeta - K_2 K_u^{-1}(F_1 - Q_1 + K_k K_2 K_u^{-1}\xi_1) \tag{144}$$

By substituting $\hat{\zeta}$ for ζ, the biomass concentration observer is then:

$$\frac{d\hat{\zeta}}{dt} = -(D + K_2 K_u^{-1} K_k)\hat{\zeta} - K_2 K_u^{-1}(F_1 - Q_1 + K_k K_2 K_u^{-1}\xi_1) \tag{145}$$

$$\hat{X} = \hat{\zeta} + K_2 K_u^{-1}\xi_1 \tag{146}$$

The procedure proposed so far is based on the use of the proper partial model algorithm set according to the process state. The problems that remain to be solved are

(a) the detection of the proper regime;
(b) the transition between the two estimation algorithms.

The first problem is solved by looking at the values of the specific growth rate estimates. If the process is not in the corresponding regime, the specific growth rate estimate directly related to ethanol (μ_r in the oxido-reductive model and μ_e in the respirative model) will have negative value. The criterion for the transition between partial model algorithm sets is given by the transition between positive and negative values of these estimates.

The transition procedure between the two sets of algorithms is more complex. In each algorithm set, five variables have to be monitored in time: the transformed variable estimate in the biomass observer ($\hat{\zeta}$), the two transformed measured state estimates (Ψ) and the two specific growth rate estimates. It is important to monitor the time course of these variables also when the estimation algorithm does not correspond to the process regime to obtain appropriate values of these variables at transition time.

For the transformed variable in the biomass observer, the following technique is suggested. When the process regime changes, the transformed variable and the estimated biomass concentration issued from the other partial model algorithm set are used to provide a value of $\hat{Z}$ for the new partial model. Then the biomass observer algorithm of the new partial model is used. As an example, if the transition between the oxido-reductive model and the respirative model occurs, then:

$$\hat{X}_{RF} = \hat{\zeta}_{RF} - K_2 K_{uRF}^{-1} \xi_{1RF} \tag{147}$$

$$\hat{\zeta}_R = \hat{X}_{RF} + K_2 K_{uR}^{-1} \xi_{1R} \tag{148}$$

For the specific growth rate estimates, distinct treatment is needed, depending whether the specific growth rate is related to ethanol or not. In the case of the estimation of μ_o, the time trajectory is still followed during the period when the partial model algorithm set is not used. Experiences have shown that the estimation error is small and that convergence following the transition is so rapid that a more sophisticated treatment is not required. For ethanol related specific growth rates, μ_e or μ_r, the estimated value is forced to zero when the partial model does not correspond to the process state.

For the transformed state variables (Ψ) of the specific growth rate estimator, the time trajectories have also to be followed during the partial model and process mismatch, but the estimated value for the biomass concentration used during this period is the one issued from the valid partial model observer. Also, the zero value of the ethanol-related specific growth rate estimate is used in the prediction equations. This technique is required to avoid a too strong perturbation of the estimation algorithm at the time of transition.

5.3.1 Experimental Verification

The growth experiments have been achieved in a 20 l BioEngineering reactor. The yeast strain, *Saccharomyces cerevisiae* and the carbon source (a mix of cane and beat molasses) were supplied by an industrial bakers' yeast producer. The operating conditions were chosen to reproduce the industrial processes. They were carried out under ethanol-concentration regulation with different nonlinear adaptive control laws ([48]). The ethanol regulation keeps the process near the boundary between the ethanol-production and ethanol-consumption states. Set-point changes, agitation speed and aeration rate perturbations have been applied to test the estimation algorithm. These perturbations create a diversity of process conditions and some of these can be considered as extreme conditions which do not appear in normal operation. The dissolved oxygen concentration was not controlled and the agitation speed (700 RPM) and the aeration rate ($2 \, \mathrm{l} \, \mathrm{l}^{-1} \, \mathrm{min}^{-1}$) were kept constant except at the time of perturbation. The *pH* was kept constant at 5.0. At this low *pH* there is no influence on the equilibrium of the different forms of dissolved carbon dioxide.

The ethanol concentration was measured in the exit gas using a Figaro sensor (TGS822) and calibrated as a function of the liquid ethanol concentration in the reactor. The dissolved oxygen concentration was measured using an Ingold probe. The carbon dioxide concentration was presumed to be directly proportional to the carbon dioxide content of the exit gas. Oxygen transfer rate (ΔQ_{O_2}) and carbon dioxide transfer rate (Q_1) were evaluated with off-gas analysis performed by a magnetic sector mass spectrometer (VG-MM8-80). The molasses feed rate was controlled by a variable speed peristaltic pump (Watson-Marlow 501U/R).

Figure 6a shows the biomass concentration estimates produced by the alternating use of the two sets of algorithms issued from the two partial models. The comparison of this time profile with the measured biomass concentration values from different samples is also shown. The precision of the biomass concentration estimate is within the precision of the measured values.

Figure 6b shows the estimated values of the specific growth rate associated with the sugar oxidation (μ_o). The estimate of μ_o remains quite constant during the non-limiting oxygen transfer condition ($t \in [1, 2]$ h) and corresponds to the maximum specific growth rate achievable without ethanol production. At $t = 2$ h the ethanol set-point was increased from $0.28 \, \text{g} \, l^{-1}$ to $1.88 \, \text{g} \, l^{-1}$. The estimate of μ_r increased accordingly. The set-point was brought back to $0.26 \, \text{g} \, l^{-1}$ at $t = 3.3$ h. The feed was decreased, resulting in a decrease of μ_o and in an increase of μ_e as expected. The agitation speed was decreased from 700 rpm to 600 rpm at $t = 4.8$ h. The process went into oxygen transfer limitation as confirmed by the sharp decrease in μ_o and in the specific oxygen uptake rate (not shown). In this case, the sugar uptake saturates the oxidation capacity of the yeast and the overflow of sugar is directed to the anaerobic pathway with ethanol production. When it was below, the sugar did not use all the oxidation capacity and allowed the ethanol oxidation to take place. This is confirmed by the values of the estimates of the two other specific growth rates (μ_e and μ_r), presented in Figs. 6c and 6d. This behaviour corresponds quite well with the hypothesis of the limiting oxidation capacity of the yeast [15]. The agitation speed was returned to 700 rpm at $t = 6.4$ h resulting in an increase in μ_o. The air flowrate was decreased at $t = 8.2$ h and increased to its previous value at $t = 9.0$ h. The estimate of μ_o decreased and increased accordingly.

The estimated values of the three specific growth rates show rapid fluctuations. To determine whether these fluctuations actually correspond to process fluctuations or are simply artefacts of the overall estimation procedure, a criterion to evaluate the accuracy of the three specific growth is proposed. The criterion is to compare the estimated respiratory quotient ($\widehat{RQ}$) obtained from the estimates of the three specific growth rates:

$$\widehat{RQ} = \frac{k_5 \hat{\mu}_o + k_6 \hat{\mu}_r + k_7 \hat{\mu}_e}{k_3 \hat{\mu}_o + k_4 \hat{\mu}_e} \tag{149}$$

to the value obtained, from experimental data. The values of each of the three specific growth rate estimates have to be accurate in order to produce a good

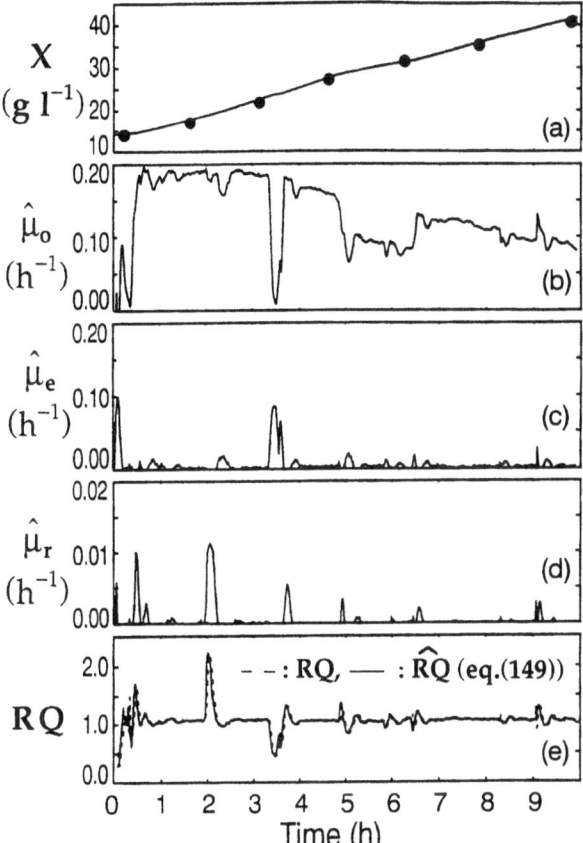

Fig. 6a–e. Estimation of biomass and specific growth rates for the bakers' yeast example

estimate of the respiratory quotient. Figure 6e shows the comparison between the experimental and the estimated respiratory quotient. A respiratory quotient over 1.06 indicates an ethanol production state and below 1.06 an ethanol consumption state. The agreement between the two curves is very good in spite of the inherent lag of the estimation procedure. This comparison allows us to have confidence in the proposed estimation procedure.

6 Adaptive Linearizing Control of Bioprocesses

6.1 Design of the Adaptive Linearizing Controller

We shall now concentrate on the design of model-based controllers for bioreactors. The key idea of the control design here is again to take advantage of what is well known about the dynamics of bioprocesses (reaction network and mass balances) which are summarized in the General Dynamical Model (10) while taking into account the model uncertainty (mainly the kinetics). Since the model is generally nonlinear, the model-based control design will result in a *linearizing* control structure, in which the on-line estimation of the unknown variables (component concentrations) and parameters (reaction rates and yield coefficients) are incorporated, the resulting controller will be an *adaptive linearizing controller*. After having introduced the general control design formulation, we shall present two typical adaptive linearizing control designs:

1. control of anaerobic digestion processes;
2. control of activated sludge processes.

One interesting aspect of the control design in the first example will be to show how to eliminate the (unknown) kinetics terms by incorporating the gas measurements into the controller. The first example is concerned with the SISO (single input–single output) control case, while the second one will concentrate on the design of a MIMO (multi-input–multi-output) adaptive linearizing controller. The design of the control algorithm is based on the General Dynamical Model (10) or on a reduced-order form of (10). In the following, we shall present two control design examples: one based on the dynamical model (10) (activated sludge process), and one based on a reduced-order version of (10) (anaerobic digestion). In each case, the control design follows the same line of reasoning described below. Let us first define the control problem:

The objective is to control the concentration of some reactant components (one in example 1, two in example 2) by acting on flow rates (the air flow rate and the recycle flow rate in example 2, the dilution rate in example 1) under the following practical constraints:

1. the components to be controlled are assumed to be measured on-line;
2. the concentrations of the other components (particularly of the biomass) are not available for on-line measurement;
3. the reaction rate vector r is unknown;
4. most of the yield coefficients are unknown (only the ratio k_1/k_2 will be assumed to be known, in example 2);
5. the mass transfer coefficients are known;
6. the feedrates F, the dilution rate D and the gaseous outflow rates Q are known (either by user's choice or by measurement).

By defining y as the controlled component(s), the dynamics of y are simply the equation(s) of y in model (10) and can be rewritten as follows:

$$\frac{dy}{dt} = -Dy + K_y r + F_y - Q_y \qquad (150)$$

where the index y holds for the rows of K, F and Q corresponding to the controlled output y. By considering the above control problem and defining

u the control input, the output Eq. (150) can be rewritten as follows:

$$\frac{dy}{dt} = f(F_y, Q_y, K_y r) + g(F_{y,y})u \tag{151}$$

Assume now that we wish to have a linear stable closed-loop (process + controller) dynamical behaviour, i.e.:

$$\frac{dy}{dt} = C_1 (y^* - y), \qquad C_1 > 0 \tag{152}$$

with y^* the desired value of y. By combining Eqs. (151) and (152), we readily obtain the control law:

$$u = g(F_y, y)^{-1} [C_1 (y^* - y) - f(F_y, Q_y, K_y r)] \tag{153}$$

Since the kinetics and most of the yield coefficients are assumed to be unknown, they are replaced by on-line estimates of selected parameters:

$$u = g(F_y, y)^{-1} [C_1 (y^* - y) - f(F_y, Q_y, \hat{K}_y, \hat{r})] \tag{154}$$

The above controller (154) is also known as the model reference adaptive linearizing control law (see e.g. [1]). As we shall see in the examples, the unknown parameters appear linearly in the equations and will therefore be estimated by using linear regression techniques (example 2) or via a Lyapunov design estimation approach (example 1). Moreover, the unknown components that may appear explicitly in the output Eq. (150) via the reaction rate r will be replaced either by an auxiliary variable, easy to compute (example 2), or by gaseous outflow rates (example 1); in the latter instance, the effect will also make the kinetics terms disappear.

6.2 Example 1: Anaerobic Digestion

The importance of implementing efficient control systems for anaerobic digestion processes clearly appears from the following two points:

1) Anaerobic digestion is intrinsically a very unstable process: variations of the input variables (hydraulic flow rate, influent organic load) may easily lead the process to a washout, i.e. a state where the bacterial life has disappeared. This phenomenon takes place under the form of acid accumulation in the reactor (see [49], [50]). It is therefore essential to implement controllers which are capable of stabilizing the process via a carefully designed control strategy.

2) If the process is used for wastewater treatment purposes, the control objective consists of maintaining the output pollution at a prescribed level despite the fluctuations of the input pollution (organic load).

From the above two comments, it is clear that control strategies should particularly concentrate on the control of the substrate concentration (which characterize the pollution level and the presence of acids). However, intricate difficulties inherent in the process make the control problem very hard to solve. Anaerobic digestion is a very complex process in which many different bacterial populations intervene. Its kinetics are basically nonlinear and non-stationary, and they are far from being fully understood. Moreover the concentrations of the different bacterial populations are not available from any direct measurement, even from off-line analyses.

Finally there remains the problem of choosing an appropriate substrate to be controlled. COD (Chemical Oxygen Demand) is a priori a very good candidate: an adaptive linearizing controller has been designed, theoretically analyzed and experimentally validated on a pilot anaerobic digester (for further details, see [51], [41], [52]). But the industrial applicability of this control solution may appear to be limited by the need of on-line COD (or equivalent substrate concentration) measurements.

Therefore there is a clear incentive to look for alternative substrate candidates. In this context, the use of hydrogen as a controlled variable appears to be very promising. As it has been recently emphasized [53], hydrogen plays an important role in the kinetics and stability of anaerobic digestion, particularly when the organic substrate is mainly composed of glucose (e.g. waste from the sugar industry). Finally hydrogen is easy to measure on-line [53].

Let us now deal with the controller design. One specific aspect of the anaerobic digestion example is that the design of the controller is based on a "reduced-order" form of the general dynamical model [(15) and (16)]. The order reduction has been done in Sect. 3.2.3 and resulted in Eq. (59) for the dissolved oxygen concentration:

$$\frac{dS_4}{dt} = -DS_4 - Q_1 - k_1 Q_3 + k_2 D S_{in} \tag{155}$$

Let us further encourage the use of a reduced order model. The design of a controller based on a reduced order model will result in a simpler controller structure but will still contain the important process features if the model reduction has been correctly performed. (Here the controller will incorporate the influent organic matter concentration S_{in} which can be viewed as a feedforward term and the methane gas production rate Q_3 which gives important information about the state of the process.) Recall that the order reduction is based on the notion of fast and slow reactions. It is difficult to know a priori which reactions in the anaerobic digestion reaction network will be, in a broad range of operating conditions, fast or slow. On the other hand, one may wish to avoid undesirable effects such as accumulation of propionate or hydrogen: then, in that situation, the corresponding reaction (reactions 2 or 4) can be considered as the limiting, i.e. "slow", reaction, and the others as "fast" reactions. For instance, assume that the main problem is to avoid accumulation of hydrogen, then reaction 4 is the slow reaction and reactions 1, 2 and 3 are the fast reactions.

Let us go back to our control problem. If we consider that the control input is the dilution rate D, the expressions of f and g are here equal to:

$$f = -Q_1 - k_1 Q_3, \; g = k_2 S_{in} - S_4 \tag{156}$$

and the controller (154) is written here as follows:

$$D = \frac{C_1 \, (S_4^* - S_4) + Q_1 + \hat{k}_1 Q_3}{\hat{k}_2 \, S_{in} - S_4} \tag{157}$$

The unknown parameters are now k_1 and k_2. They can be estimated on-line by using an appropriate updating, for instance here a "Lyapunov design" estimation equation:

$$\frac{d\hat{k}_i}{dt} = C_{2i} \, (S_4^* - S_4), \qquad C_{2i} > 0, \quad i = 1, 2 \tag{158}$$

A typical simulation is shown in Fig. 7 (see also [54], [55] [56]). In this simulation, the hydrogen gaseous outflow rate Q_1 has been assumed to be negligible, and the parameter k_2 has been assumed to be known. Therefore, only k_1 has been estimated on-line. The anaerobic process has been simulated by integration of the mass balance equations with the following yield coefficients, specific growth rate expressions, and initial conditions:

Yield coefficients

$$k_1 = 3.2, \quad k_2 = 1.5, \quad k_3 = 0.7, \quad k_4 = 12, \quad k_5 = 0.27, \quad k_6 = 0.53 \tag{159}$$

$$k_7 = 1.15, \quad k_8 = 0.6, \quad k_9 = 0.1, \quad k_{14} = 0.3, \quad k_{15} = 0.08 \tag{160}$$

Specific growth-rate expressions

$$\mu_1 = \frac{\mu_{max1} \, S_1}{1 + (S_2 + S_3)/K_{I_1}}, \qquad \mu_2 = \frac{\mu_{max2} \, S_2}{(K_{S_2} + S_2)(1 + S_3/K_{I_2})} \tag{161}$$

$$\mu_3 = \frac{\mu_{max3} \, S_3}{K_{S_3} + S_3 + S_3^2/K_{I_3}}, \qquad \mu_4 = \frac{\mu_{max4} \, S_4}{K_{S_4} + S_4 + S_4^2/K_{I_4}} \tag{162}$$

with:

$$\mu_{max1} = 0.2, \quad K_{I_1} = 10, \quad \mu_{max2} = 0.5, \quad K_{S_2} = 0.4, \quad K_{I_2} = 0.5 \tag{163}$$

$$\mu_{max3} = 0.4, \quad K_{S_3} = 0.5, \quad K_{I_3} = 4, \quad \mu_{max4} = 0.5, \quad K_{S_2} = 4, \quad K_{I_4} = 3 \tag{164}$$

Initial steady-state conditions

$$S_1(0) = 0.79 \text{ g l}^{-1}, \quad S_2(0) = 0.28 \text{ g l}^{-1},$$

$$S_3(0) = 0.31 \text{ g l}^{-1}, \quad S_4(0) = 3.0 \, \mu M \tag{165}$$

$$X_1(0) = 7.6 \text{ g l}^{-1}, \quad X_2(0) = 3.3 \text{ g l}^{-1},$$

$$X_3(0) = 0.29 \text{ g l}^{-1}, \quad X_4(0) = 1.6 \text{ g l}^{-1} \tag{166}$$

The performance of the controller was evaluated for a change in the influent substrate concentration (S_{in1}) from 25 to 35 g l^{-1}, applied at $t = 2$ d. The value of the design parameter C_1 was chosen to be equal to $1\ d^{-1}$. The parameter estimation dynamics is imposed by C_2. As for the on-line estimation of the specific growth rates (Sect. 5), in order to have a closed-loop dynamics independent of the (time-varying) value of the regressor Q_3, the design parameter C_2 was chosen to be inversely proportional to the regressor, i.e.:

$$C_2 = \frac{C_1}{4Q_3^{\frac{2}{3}}} \tag{167}$$

The concentration of each substrate and bacterial population is shown in Fig. 7a and 7b, respectively, for dissolved hydrogen control. The controller proves to be able to maintain the hydrogen concentration at the desired set-point. The concentration of bacterial populations slowly reaches new levels while the other substrates go back to their initial values.

Remark: A similar approach has been used to design an adaptive linearizing controller for propionate as well as for COD (As already mentioned). It is worth noting that the three control designs result in similar controller structures ([54], [55]). The controllers of propionate and of COD have been successfully applied to a pilot-scale anaerobic digestion reactor ([52], [57]).

Discussion

One of the key feature of the adaptive linearizing can be illustrated on the basis of the anaerobic digestion example: it allows the well-known characteristics of the process dynamics to be incorporated, while keeping the usual features of classical controllers:

(a) proportional action: via the term $C_1(S_4^* - S_4)$;
(b) integral action: via the adaptation mechanism (158);

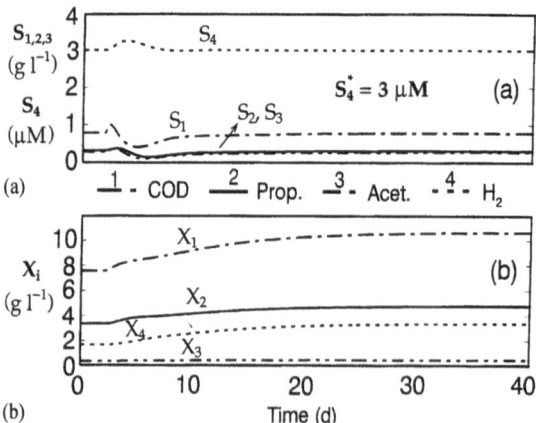

Fig. 7a, b. Control of the hydrogen concentration in an anaerobic digestion reactor

(c) feedforward action: via the presence of S_{in}. Indeed the controller is capable of anticipating the effect of a variation of the influent substrate concentration: for instance, an increase of S_{in} will result in a decrease of the control input D, inversely proportional to this increase.

Besides the controller contains a state "estimate" via the term $\hat{k}_1 Q_3$. If the process is in good working conditions. In particular if the biomass is in a good state, then the process produces a lot of methane. Then it is possible to treat high amounts of organic matter, since the control action D is proportional to Q_3.

Finally note that beside the integral action, the estimation of "physical" parameters (here k_1, which is indeed a conversion coefficient of hydrogen into methane) has the further advantage of giving useful information, which can be used for monitoring the process, and possibly also for analyzing the internal working of the process.

6.3 Example 2: Activated Sludge Process

The following example is treated in detail in [55]. The control problem we shall now consider can be formulated as follows, in any aerobic process, proper aeration is crucial to process efficiency, and adequate control of the dissolved oxygen concentration in the aerator is very important. But it is also important to limit load variations and substrate concentration variations by acting on them. As suggested by many authors (e.g. [17], [58]), load variations can be expected to be smoothed by using the recycle flow rate as a control input. We shall consider the above control algorithm under the following conditions:

1. The controlled outputs are the effluent BOD (Biological Oxygen Demand) and the dissolved oxygen concentrations, which are assumed to be measured on-line (see [59] for an on-line measuring device of the BOD in an activated sludge processes).
2. The control inputs are the recycle flow rate and the air flow rate.

Therefore the input u and the output y are defined as follows:

$$u = \begin{pmatrix} F_R \\ W \end{pmatrix}, \qquad y = \begin{pmatrix} S \\ C \end{pmatrix} \tag{168}$$

and the control problem is multivariable (2 inputs, 2 outputs). The expressions of f and g in (151) are readily obtained from model (27):

$$f = \begin{pmatrix} -k_1 \mu X + D_{in}(S_{in} - S) \\ -k_2 \mu X - D_{in} C \end{pmatrix} \qquad g = \begin{pmatrix} -\frac{S}{V} & 0 \\ -\frac{C}{V} & a_o(C_S - C) \end{pmatrix} \tag{169}$$

g is invertible (i.e. g^{-1} exists) as long as S and $C_S - C$ are different from zero. Under these physically realistic conditions, the linearizing controller (154) can be applied. The remaining problem is how to deal with the "unknown" parameters k_1, k_2 and μ, and variable X. The solution proceeds as follows. Let us first define the auxiliary variables ζ (as introduced in Sect. 3.1).

$$z = \begin{pmatrix} \zeta_1 \\ \zeta_2 \\ \zeta_3 \end{pmatrix} = \begin{pmatrix} k_1 \\ k_2 \\ 0 \end{pmatrix} X + \begin{pmatrix} 1 & 0 & 0 \\ 0 & 1 & 0 \\ 0 & 0 & k_2 \end{pmatrix} \begin{pmatrix} S \\ C \\ X_R \end{pmatrix} \tag{170}$$

The dynamical equation of ζ are derived from model (27) and definitions (170):

$$\frac{d}{dt} \begin{pmatrix} \zeta_1 \\ \zeta_2 \\ \zeta_3 \end{pmatrix} = \begin{pmatrix} -D_1 & 0 & D_2 k_1/k_2 \\ 0 & -D_1 & D_2 \\ 0 & D_3 & -D_4 \end{pmatrix} \begin{pmatrix} \zeta_1 \\ \zeta_2 \\ \zeta_3 \end{pmatrix} + \begin{pmatrix} D_{in} S_{in} \\ \Delta Q_{O_2} \\ -D_3 C \end{pmatrix} \tag{171}$$

with D_i ($i = 1$ to 4) as defined in Sect. 2.5. One important feature of the above dynamics (171) is that these are independent of the (unknown) kinetics. Equation (171) is a dynamical system linear-in-the-state ζ. We can check that its state matrix is asymptotically stable if $F_{in} - F_R > 0$ and $F_W > 0$. Therefore Eq. (171) can be used to compute the value of ζ on-line from the knowledge of the flow rates F_{in}, F_R and F_W, the flow rates $F_{in} D_{in}$ and ΔQ_{O_2}, the dissolved oxygen concentration C and the ratio k_1/k_2. Let us now rewrite the specific growth rate μ as follows:

$$\mu = \alpha S C \tag{172}$$

where α is an (unknown) positive function of the process components: Eq. (172) simply implies that there is no growth in absence of one of the limiting substrates. By introducing (170) and (172), the function f in (169) becomes:

$$f = \begin{pmatrix} -\alpha S C(\zeta_1 - S) + D_{in}(S_{in} - S) \\ -\alpha S C(\zeta_2 - C) - D_{in} C \end{pmatrix} \tag{173}$$

This expression of f is used in the computation of the control law (154):

$$F_R = -\frac{V}{S}[C_{11}(S* - S) - D_{in}(S_{in} - S) + \hat{\alpha}_1 S C(\zeta_1 - S)] \tag{174}$$

$$W = \frac{1}{a_0(C_S - C)}\left[C_{12}(C* - C) - \frac{C_{11}(S* - S)C}{S} + \frac{D_{in} S_{in} C}{S} \right.$$

$$\left. + \hat{\alpha}_2 C(S\zeta_2 - C\zeta_1) \right] \tag{175}$$

C_{11} and C_{12} (>0) are the control design parameters (we have considered here a decoupling diagonal matrix C_1). The variables ζ_1 and ζ_2 are computed via the use of Eq. (171) (which only requires a knowledge of the ratio k_1/k_2) and the parameter α is estimated on-line by using e.g. a recursive least square (RLS) algorithm. In discrete-time, if α is estimated via the substrate concentration equation S [α_1, to be used in Eq. (174)] or via the dissolved oxygen concentration C [α_2, to be used in Eq. (175)], the RLS algorithm is written as follows:

$$\hat{\theta}_{i,t+1} = \hat{\theta}_{i,t} + g_{i,t} \phi_{i,t} e_{i,t+1} \qquad i = 1, 2 \tag{176}$$

$$g_{i,t} = \frac{g_{i,t-1}}{\gamma_i + g_{i,t-1}\,\phi_{i,t}^2} \qquad\qquad 0 < \gamma_i \le 1 \tag{177}$$

with:

$$\hat{\theta}_{1,t} = \hat{\alpha}_{1,t}, \quad \phi_{1,t} = TS_t\,C_t(S_t - \zeta_{1,t}) \tag{178}$$

$$e_{1,t+1} = S_{t+1} - S_t + TD_{1,t}\,S_t - TD_{in,t}\,S_{in,t} - \hat{\alpha}_{1,t}\,\phi_{1,t} \tag{179}$$

$$\hat{\theta}_{2,t} = \hat{\alpha}_{2,t}, \quad \phi_{2,t} = TS_t\,C_t(C_t - \zeta_{2,t}) \tag{180}$$

$$e_{2,t+1} = C_{t+1} - C_t + TD_{1,t}\,C_t - Ta_0W_t(C_s - C_t) - \hat{\alpha}_{2,t}\,\phi_{2,t} \tag{181}$$

and t the time index, and γ_i ($i = 1, 2$) a forgetting factor. Note that now because of the variables ζ, the biomass X does not appear explicitly anymore in the controller (174) and (175).

A typical simulation result is shown in Fig. 8 (see also [55]). The activated sludge process has been simulated by numerical integration of the basic dynamical model equation (27) using a 4th order Runge-Kutta method) with the following (Monod-type) model for the specific growth rate:

$$\mu = \mu_{max}\frac{S}{K_S + S}\frac{C}{K_C + C} \tag{182}$$

and a decay rate $(-k_dX)$ which has been added to the biomass equation (in order to simulate biomass mortality). The specific growth rate model and the decay rate term are (obviously) completely ignored by the control algorithm. The decay rate term can be viewed as an unmeasured disturbance. The model parameters have been set to the following values (inspired from literature data, in particular from [58], [60]):

$$k_1 = 1.2, \quad k_2 = 0.565, \quad \mu_{max} = 0.2\,\mathrm{h}^{-1},$$

$$K_S = 75\,\mathrm{mg\,l}^{-1}, \quad k_d = 0.001\,\mathrm{h}^{-1}$$

$$a_0 = 0.018\,\mathrm{m}^{-3}, \quad C_S = 10\,\mathrm{mg\,l}^{-1},$$

$$V = 100\,\mathrm{m}^3, \quad V_S = 50\,\mathrm{m}^3, \quad K_C = 2\,\mathrm{mg\,l}^{-1}$$

In Fig. 8, a square wave signal of the influent BOD concentration (disturbance input) S_{in} (from 150 mg l^{-1} to 200 mg l^{-1}) (in order to simulate the periodical variation of the pollutant load) has been applied over period of 20 d (480 h). The sampling period has been set to 3 min following the constraints of the commercially available BOD measuring device proposed by [59]. The following initial process conditions have been considered in the simulation:

$$S = 5\,\mathrm{mg\,l}^{-1}, \quad C = 6\,\mathrm{mg\,l}^{-1},$$

$$X = 1225\,\mathrm{mg\,l}^{-1}, \quad X_R = 2333\,\mathrm{mg\,l}^{-1} \tag{183}$$

The controller parameters have been set to the following values:

$$S^* = 5\,\mathrm{mg\,l}^{-1}, \quad C^* = 6\,\mathrm{mg\,l}^{-1},$$

$$C_{11} = 1\,\mathrm{h}^{-1}, \quad C_{12} = 10\,\mathrm{h}^{-1} \tag{184}$$

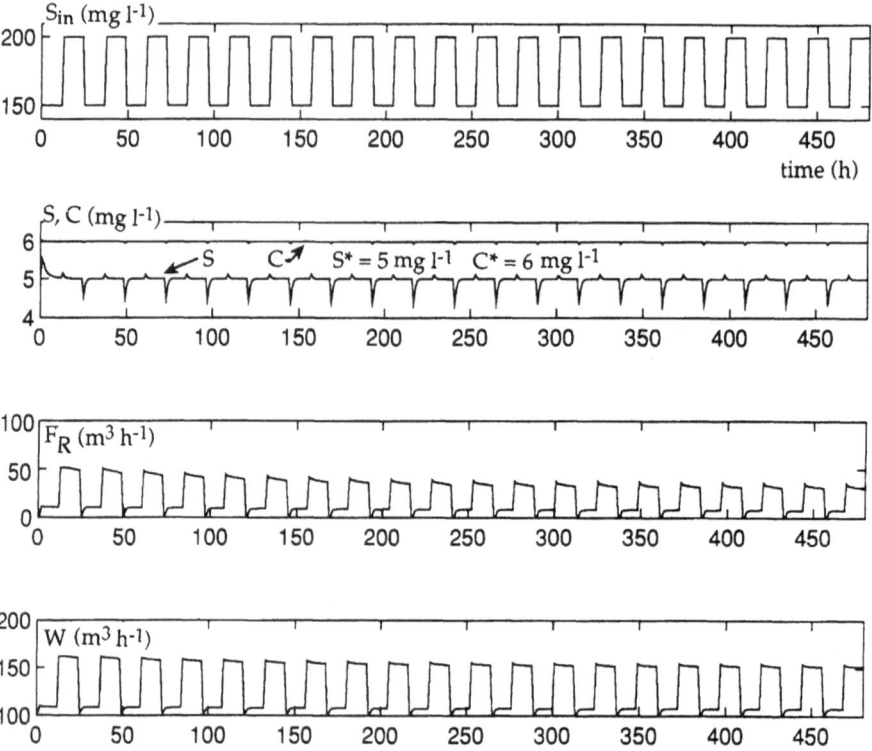

Fig. 8. Control of an activated sludge process in the presence of an unknown biomass mortality

The auxiliary variables ζ, the unknown parameters and tuning estimation variables have been initialized as follows:

$$\zeta_1 = 1400 \text{ mg l}^{-1}, \quad \zeta_2 = 750 \text{ mg l}^{-1}, \quad \zeta_3 = 1400 \text{ mg l}^{-1},$$

$$g_{1,0} = g_{2,0} = 10^{-3}, \quad \gamma_1 = \gamma_2 = 0.9,$$

$$\hat{\alpha}_{1,0} = \hat{\alpha}_{2,0} = 0.00025 \text{ l}^2 \text{ mg}^{-2} \text{ h}^{-1}$$

Note the ability of the controller to maintain the controlled outputs S and C close to their desired values in spite of the unknown disturbance.

7 Conclusions

The objective of this paper was to present a survey of recent approaches to model-based monitoring and control of bioreactors. The proposed results cover the whole range from theory (dynamical modelling, dynamical analysis, monitoring and control design) to practice (experimental results). They apply to different types of reactors, mainly stirred tank reactors and fixed bed reactors, but could also be extended, for example, to fluidised bed reactors. The dynamical model of the bioprocesses is based on material balances and is formalized into a General Dynamical Model framework which serves as a basis for dynamical analysis, and for monitoring and control algorithm design. Monitoring was found to be a key question in bioprocess applications. The design of asymptotic observers for the concentrations of the process components has been developed, analyzed and illustrated. Due to the usually large uncertainty of some process parameters, such as the process kinetics, on-line estimation of the uncertain parameters has been considered either for monitoring purposes or to be included in an adaptive model-based control scheme. In the latter case, the incorporation of "physical" parameters presents the double advantage of introducing an integral action into the controller while giving extra information about the process behaviour and performance.

Acknowledgements. This paper presents research results of the Belgian Programme on Interuniversity Poles of Attraction initiated by the Belgian State, Prime Minister's Office, Science, Technology and Culture. The scientific responsibility rests with its authors.

References

1. Bastin G, Dochain D (1990) On-line Estimation and Adaptive Control of Bioreactors. Elsevier, Amsterdam
2. Chen L, Bastin G, Van Breusegem V (1991) Adaptive nonlinear regulation of fed-batch biological reactors: an industrial application. Proc. 30th IEEE CDC, p. 2130
3. Henson MA, Seborg D (1992) Chem Eng Sci 47(4): 821
4. Flaus JM, Cheruy A, Engasser JM (1991) J Proc Cont 1: 271
5. Dahhou B, Bordeneuve J, Babary JP (1991) Multivariable long-range predictive control algorithm applied to a continuous flow fermentatin process. Proc. World IFAC Congress, p. 393
6. Golden MP, Pangrie BJ, Ydstie BE (1986) Nonlinear adaptive optimization of a continuous bioreactor. Proc. AIChE 1986 National Meeting, Pap. 125b
7. Alvarez JG, Alvarez JG (1988) Proc ACC 2: 1112
8. Hoo KA, Kantor JC (1986) Chem Eng Comm 46: 38
9. De Tremblay M, Perrier M (1992) Bioprocess Eng 7: 229
10. De Tremblay M, Perrier M, Chavarie C, Archambault J (1993) Bioprocess Eng. (in press)
11. Chotteau V, Bastin G (1992) Identification of a reaction mechanism for a class of animal cell cultures. Proc ICCAFT 5/IFAC-BIO 2, p. 215
12. Van Impe J. (1993) Modelling and Optimal Adaptive Control of Biotechnological Processes. PhD thesis, Katholieke Universiteit Leuven, Belgium
13. Van Impe J, Nicolai B, Vanrolleghem P, Spriet J, De Moor B, Van De Walle J (1992) Chem Eng Comm 117: 337
14. Mosey FE (1983) Water Sci Technol 15: 209
15. Sonnleitner B, Käppeli O (1986) Biotechnol Bioeng 28: 927

16. Hamäläinen RP, Halme A, Gyllenberg A (1975) A control model for activated sludge wastewater treatment process. Proc 6th IFAC World Congress, Boston, Paper 61: 6
17. Marsili-Libelli S (1984) Trans Inst Meas Control 6: 146
18. Holmberg A, Ranta J (1982) Automatica 18: 181
19. Danckwerts PV (1953) Chem Eng Sci 2 (1): 1
20. Feyo de Azevedo S, Romero-Ogawa MA, Wardle AP (1990) Trans I Chem E 68 (Part A): 2–8
21. Dochain D (1994) Contribution to the Analysis and Control of Distributed Parameter Systems with Application to (Bio) chemical Processes and Robotics. Thèse d'aggrégation de l'enseignement supérieur, Université Catholique de Louvain, Belgium
22. Dochain D, Babary JP, Tali-Maamar MN (1992) Automatica 68: 873
23. Dochain D, Tali-Maamar N, Babary JP (1994) Design of adaptive linearizing controllers for fixed bed reactors. Proc. ACC 1994, p. 335
24. Aoufoussi H, Perrier M, Chaouki J, Chavarie C, Dochain D (1992) Can J Chem Eng 70: 356
25. Dochain D, Bouaziz B (1994) Mathematics and Computers in Simulation. 37(2–3): 165
26. Bouaziz B, Dochain D (1993) Control analysis of fixed bed reactors: a singular perturbation approach. Proc ECC'93, 1741
27. Chen L (1992) Modelling, Identifiability and Control of Complex Biotechnological Processes. PhD thesis, Université Catholique de Louvain, Belgium
28. Gavalas GR (1968) Nonlinear Differential Equations of Chemically Reacting Systems. Springer Verlag, Berlin
29. Fjeld M. Asbjörnsen OA, Aström KJ (1974) Chem Eng Sci 29: 1917
30. Van Breusegem V, Bastin G (1993) A singular perturbation approach to the reduced order dynamical modelling of reaction systems. Submitted for publication
31. Kwakernaak H, Sivan R (1972) Linear Optimal Control Systems. John Wiley, New York
32. Stephanopoulos G, San K-Y (1984) Biotechnol Bioeng 26: 1176
33. Lee SH, Tsobanakis P, Phillips JA, Georgakis C (1992) Issues in the optimization, estimation and control of fed-batch bioreactors using tendency models. Proc ICCAFT 5/IFAC-BIO 2, Keystone, Colorado
34. Yoo YJ, Hong J, Hatch RT (1985) Proc ACC 2: 866
35. Caminal G, Lafuente FJ, Lopez-Santin J, Poch M, Sola C (1987) Biotechnol Bioeng 24: 366
36. Ljung L (1979) IEEE Trans Aut Cont 24: 36
37. Bastin G, Levine J (1990) On state reachability of reaction systems. Proc 29th CDC, p. 2819
38. Dochain D, Chen L (1992) Local observability and controllability of stirred tank reactors. J. Proc Cont 2 (3): 139
39. Narendra KS, Annaswamy AM (1989) Stable Adaptive Systems Prentice-Hall, Engelwood Cliffs, NJ
40. Marino R (1990) IEEE Trans Aut Cont 35: 1054
41. Dochain D (1986) On-line Parameter Estimation, Adaptive State Estimation and Adaptive Control of Fermentation Processes. PhD thesis, Université Catholique de Louvain, Belgium
42. Dochain D, Bastin G (1987) Convention de recherche Solvay – UCL, rapport final
43. Georgakis C, Aris R, Amundson R (1977) Chem Eng Sci 32: 1359
44. Jorgensen SB (1986) Fixed bed reactor dynamics and control – A review. Proc. IFAC Control of Distillation Columns and Chemical Reactors, Pergamon, p. 11
45. Villadsen J, Michelsen ML (1978) Solution of Differential Equation Models by Polynomial Approximation. Prentice-Hall, Engelwood Cliffs, NJ
46. Ray WH, Lainiotis DG (1978) Distributed Parameter Systems, Identification, Estimation and Control. Marcel Dekker, New York
47. Pomerleau Y, Perrier M (1990) AIChE J 36 (2): 207
48. Pomerleau Y (1990) Modélisation et Contrôle d'un Procédé Fed-batch de Culture des Levures à Pain Saccharomyces cerevisiae. PhD thesis, Ecole Polytechnique de Montréal, Canada
49. Binot R, Bol T, Naveau H, Nyns EJ (1983) Wat Sci Tech 15: 103
50. Fripiat JL, Bol T, Binot R, Naveau H, Nyns EJ (1984) A strategy for the evaluation of methane production from different types of substrate biomass. In R. Buvet, M.F. Fox and D.J. Picker (eds.), Biomethane, Production and uses, Roger Bowskill ltd, Exeter, UK: 95
51. Dochain D, Bastin G (1984) Automatica 20: 621
52. Renard P, Dochain D, Bastin G, Naveau H, Nyns EJ (1988) Biotechnol Bioeng 31: 287–294
53. Pauss A, Beauchemin C, Samson R, Guiot S (1990) Biotechnol Bioeng 35: 492
54. Perrier M, Dochain D (1993) Int J Adaptive Cont Signal Proc 7 (4): 309
55. Dochain D, Perrier M (1992) Adaptive linearizing control of activated sludge processes. Proc Control Systems'92, p. 211

56. Dochain D, Perrier M, Pauss A (1991) Ind Eng Chem Res 30: 129
57. Renard P, Van Breusegem V, Nguyen N, Naveau H, Nyns EJ (1991) Biotechnol Bioeng 38: 805
58. Holmberg A (1983) A microprocessor-based estimation and control system for the activated sludge process. In A Halme (Ed.), Modelling and Control for Biotechnical Processes, Pergamon: 111
59. Khöne M (1985) Practical experiences with a new on-line BOD measuring device. Env Technol Letters 6: 546
60. Marsili-Libelli S (1989) Adv Biochem Eng/Biotechnol 38: 90

Author Index Volume 1-56

Author Index Vols. 1-50 see Vol. 50

Bajpai, P., Bajpai, P. K.: Realities and Trends in Emzymatic Prebleaching of Kraft Pulp. Vol. 56, p. 1
Bajpai, P. K.: see Bajpai, P.: Vol. 56, p. 1
Bárzana, E.: Gas Phase Biosensors. Vol. 53, p. 1
Bazin, M. J. see Markov, S. A.: Vol. 52, p. 59
de Bont, J.A.M.: see van der Werf, M. J.: Vol. 55, p. 147

Chang, H. N. see Lee, S. Y.: Vol. 52, p. 27
Cheetham, P. S. J.: Combining the Technical Push and the Business Pull for Natural Flavours. Vol. 55, p. 1
Croteau, R: see McCaskill, D.: Vol. 55, p. 107

Dochain, D., Perrier, M.: Dynamical Modelling, Analysis, Monitoring and Control Design for Nonlinear Bioprocesses. Vol. 56, p. 147
Dutta, N. N.: see Ghosh, A. C.: Vol. 56, p. 111

Eggeling, L., Sahm, H., de Graaf, A.A.: Quantifying and Directing Metabolite Flux: Application to Amino Acid Overproduction. Vol. 54, p.1
Ehrlich, H. L. see Rusin, P.: Vol. 52, p. 1

Fiechter, A. see Ochsner, U. A.: Vol. 53, p. 89
Freitag, R., Hórvath, C.: Chromatography in the Downstream Processing of Biotechnological Products. Vol. 53, p. 17

Gatfield, I.L.: Biotechnological Production of Flavour-Active Lactones. Vol. 55, p.221
Ghosh, A. C., Mathur, R. K., Dutta, N. N.: Extraction and Purification of Cephalosporin Antibiotics. Vol. 56, p. 111
Ghosh, P. see Singh, A.: Vol. 51, p. 47
de Graaf, A.A. see Eggeling, L.: Vol. 54, p.1
de Graaf, A.A. see Weuster-Botz, D.: Vol. 54, p. 75
de Graaf, A.A. see Wiechert, W.: Vol. 54, p.109
Gros, J.-B.: see Larroche, C.: Vol. 55, p. 179
Gutman, A. L., Shapira, M.: Synthetic Applications of Enzymatic Reactions in Organic Solvents. Vol. 52, p. 87

Hall, D. O. see Markov, S. A.: Vol. 52, p. 59
Hasegawa, S., Shimizu, K.: Noninferior Periodic Operation of Bioreactor Systems. Vol. 51, p. 91
Hembach, T. see Ochsner, U. A.: Vol. 53, p. 89
Hiroto, M. see Inada, Y.: Vol. 52, p. 129

Hórvath, C. see Freitag, R.: Vol. 53, p. 17

Inada, Y., Matsushima, A., Hiroto, M., Nishimura, H., Kodera, Y.: Chemical Modifications of Proteins with Polyethylen Glycols. Vol. 52, p. 129

Johnson, E. A., Schroeder, W. A.: Microbial Carotenoids. Vol. 53, p. 119

Kawai, F.: Breakdown of Plastics and Polymers by Microorganisms. Vol. 52, p. 151
Kodera, F. see Inada, Y.: Vol. 52, p. 129
Krämer, R.: Analysis and Modeling of Substrate Uptake and Product Release by Procaryotic and Eucaryotik Cells. Vol. 54, p. 31
Kuhad, R. Ch. see Singh, A.: Vol. 51, p. 47

Larroche, C. , Gros, J.-B.: Special Transformation Processes Using Fungal Spares and Immobilized Cells. Vol. 55, p. 179
Leak, D. J.: see van der Werf, M. J.: Vol. 55, p. 147
Lee, S. Y., Chang, H. N.: Production of Poly(hydroxyalkanoic Acid). Vol. 52, p. 27
Lievense, L. C., van 't Riet, K.: Convective Drying of Bacteria II. Factors Influencing Survival. Vol. 51, p. 71

Markov, S. A., Bazin, M. J., Hall, D. O.: The Potential of Using Cyanobacteria in Photobioreactors for Hydrogen Production. Vol. 52, p. 59
Mathur, R. K.: see Ghosh, A. C.: Vol. 56, p. 111
Matsushima, A. see Inada, Y.: Vol. 52, p. 129
McCaskill, D., Croteau, R.: Prospects for the Bioengineering of Isoprenoid Biosynthesis. Vol. 55, p. 107
McLoughlin, A. J.: Controlled Release of Immobilized Cells as a Strategy to Regulate Ecological Competence of Inocula. Vol. 51, p. 1
Mukhopadhyay, A.: Inclusion Bodies and Purification of Proteins in Biologically Active Forms. Vol. 56, p. 61

Nishimura, H. see Inada, Y.: Vol. 52, p. 123

Ochsner, U. A., Hembach, T., Fiechter, A.: Produktion of Rhamnolipid Biosurfactants. Vol. 53, p. 89

Perrier, M.: see Dochain, D.: Vol. 56, p. 147

van 't Riet, K. see Lievense, L. C.: Vol. 51, p. 71
Roychoudhury, P. K., Srivastava, A., Sahai, V.: Extractive Bioconversion of Lactic Acid. Vol. 53, p. 61
Rusin, P., Ehrlich, H. L.: Developments in Microbial Leaching – Mechanisms of Manganese Solubilization. Vol. 52, p. 1
Rogers, P. L., Shin, H. S., Wang, B.: Biotransformation for L-Ephedrine Production. Vol. 56, p. 33

Sahai, V. see Singh, A.: Vol. 51, p. 47
Sahai, V. see Roychoudhury, P. K.: Vol. 53, p. 61
Sahm, H. see Eggeling, L.: Vol. 54, p. 1

Schreier, P.: Enzymes and Flavour Biotechnology. Vol. 55, p. 51

Schroeder, W. A. see Johnson, E. A.: Vol. 53, p. 119

Scouroumounis, G. K.: see Winterhalter, P.: Vol. 55, p. 73

Scragg, A.H.: The Production of Aromas by Plant Cell Cultures. Vol. 55, p. 239

Shapira, M. see Gutman, A. L.: Vol 52, p. 87

Shimizu, K. see Hasegawa, S.: Vol. 51, p. 91

Shin, H. S.: see Rogers, P. L., Vol. 56, p. 33

Singh, A., Kuhad, R. Ch., Sahai, V., Ghosh, P.: Evaluation of Biomass. Vol. 51, p. 47

Sonnleitner, B.: New Concepts for Quantitative Bioprocess Research and Development. Vol. 54, p. 155

Srivastava, A. see Roychoudhury, P. K.: Vol. 53, p. 61

Wang, B.: see Rogers, P. L., Vol. 56, p. 33

van der Werf, M. J., de Bont, J. A. M. Leak, D. J.: Opportunities in Microbial Biotransformation of Monoterpenes. Vol. 55, p. 147

Weuster-Botz, D., de Graaf, A.A.: Reaction Engineering Methods to Study Intracellular Metabolite Concentrations. Vol. 54, p. 75

Wiechert, W., de Graaf, A.A.: In Vivo Stationary Flux Analysis by ^{13}C-Labeling Experiments. Vol. 54, p. 109

Wiesmann, U.: Biological Nitrogen Removal from Wastewater. Vol. 51, p. 113

Winterhalter, P., Skouroumounis, G. K.: Glycoconjugated Aroma Compounds: Occurence, Role and Biotechnological Transformation. Vol. 55, p. 73

Subject Index

Activated sludge process 158
Adaptive linearizing control law 187
Adaptive linearizing controller 186
Adsorptive separation 118
Aggregation 65
Anaerobic digestion 155, 164
Anion exchange chromatography 94
Antimicrobial activity 117
Approximation methods 177
Aqueous two-phase systems (ATPS)
 126
Asymptotic observers 166
Axial diffusion 159

Bamboo kraft pulp 16
Benzaldehyde concentration 37
Benzaldehyde feeding 42
Benzyl alcohol production 37
Bioassay 98
Biobleaching 7
Biodegrable polymer 152
Biological bleaching 5
Biologically active conformation 71
Biomass observer 183
Biotransformation 36
Bleaching sequence 11
Brown stock black liquor 11
Brown white water 11
By-product inhibition 41

Centrifugation 84
Cephalosporin 112
Cephalosporin-C (CPC) 112
Cephalosporium acremonium 112
Chaotropic agents 86

Chaperone proteins 67
Characterization 96
Chlorine-free bleaching 14
Commercial xylanases 9
Conformational restriction 76
Continuous processes 42, 55
Crystallization 127

Danckwerts boundary conditions 160
Detergents 86
Diafiltration 88
Dialysis 88
Diffusional limitations 47
Dilution rate 154
Discrete-time form 170
Disintegration 83
Disulfide bond formation 73
Downstream processing 81
Dynamical equations 153
Dynamical model 157

Electrophoresis 135
Emulsion liquid membrane 122
End-product inhibition 41
Endo-β-xylanase 6
Enzyme dispersion 9
Expression level 87
Extended Kalman filter 165
Extractive esterification 121

Facilitated transport 123
Fed-batch process 41
Fed-batch systems 42
Fermentative metabolism 37
Filtration properties 116

First acidogenic path   155
Fixed bed reactors   159
Formation of inclusion body   66

Gel electrophoresis   139
Gel filtration   88, 134
General Dynamical Model   155, 158
Growth associated productions   153

Hardwood pulps   14
Heat shock proteins   67
Hemicellulolytic enzymes   7
High performance liquid chromatography
   (HPLC)   134
Hollow Fibre Contained Liquid Mem-
   brane (HFCLM)   125
Hydrogenophilic methanogenic bacteria
   156
Hydrophobic chromatography   130
Hydrophobicity   69
Hydroxyapatite chromatography   130

Immobilizing matrix   47
Impurities   87
In vitro folding   77
In vitro refolding efficiencies   91
In vivo folding   66
In vivo protein folding   69
Inclusion bodies   65
Inhibitory effects   44
Insulin   102
Integrating membranes   125
Ion-exchange chromatography   129
ISO brightness   14
Isoelectric focussing   139

Kalman observers   165
k_L 159

L-ephedrine   33
L-PAC   36
L-phenylacetylcarbinol   36
Large-scale zone electrophoresis   135
Level of purity   99
Linearizing control structure   186

Lipophilic intermediates   119
Liquid-liquid extraction   120
Luenberger observers   165
Lyapunov design estimation   187

Metabolic pathway   153

N-terminal processing   95
Non-dispersive reactive extraction   141
Non-dispersive solvent extraction   124
Non-growth associated productions   153
Non-specific precipitation   66
Nutritional requirements   113

Observer-based estimator   179
On-line knowledge   178
On-line measurements   168
Osmotic lysis   116
Oxidative folding   70
Oxygen delignification   17
Oxygen mass transfer coefficient   159

Paper chromatography   128
Pine kraft pulp   13
Plant scale trial   22
Polishing   81
Polishing steps   93
Poly-β-hydroxybutyric acid (PHB)   153
Polyacryl amide gel electrophoresis   139
Pre-clinical safety   99
Precipitation   127
Preprotein   71
Proline isomerization   72
Proteins aggregate in vitro   77
Proteolytic activity   67
Pulp bleaching   5
Pulping process   10
Purity   99
Pyruvate decarboxylase (PDC)   36
Pyruvate depletion   42

Radio-receptor binding   98
Reactive extraction   122
Reductive growth   157
Reference standards   97

Refolding   88
– by diafiltration   88
– by dialysis   88
– by dilution   89
– by gel filtration   88
– by stage dilution   90
Refractile bodies   65
Respiratory growth   157
Respiratory quotient   37
Reversed miscelles   92

Sampling period   170
Second acidogenic path   155
Separation
– of cell biomass   115
– of inclusion bodies   83
7-amonocephalosporanic acid (7-ACA)
   113
Softwood kraft pulp   21
Softwood pulps   14
Solubilization   68

– of inclusion bodies   83
Solvent extraction   120
State transformation   161
Substrate inhibition   50
Supported liquid membrane   124
Synthesizing enzyme   116

TCF bleaching   12
Thin layer chromatography   128
Tissue plasminogen activator   102
Two-dimensional electrophoresis   139

Ultrastructure studies   68

Wastewater treatment processes   152

Xylanase prebleaching   7
Xylanase pretreatment   25

Yeast growth   152, 157